★‖‖‖★

THE BLACK PRESIDENCY

THE BLACK PRESIDENCY

★|||||★

Barack Obama
and the Politics of Race in America

MICHAEL ERIC DYSON

Houghton Mifflin Harcourt
Boston   New York
2016

For information about permission to reproduce selections from
this book, write to trade.permissions@hmhco.com or to Permissions,
Houghton Mifflin Harcourt Publishing Company, 3 Park Avenue,
19th Floor, New York, New York 10016.

www.hmhco.com

Library of Congress Cataloging-in-Publication Data is available.
ISBN 978-0-544-38766-9

Book design by Lisa Diercks

Printed in the United States of America
DOC 10 9 8 7 6 5 4 3 2 1

For Marcia

★ ||||||| ★

CONTENTS

★||||||★

INTRODUCTION:
THE BURDEN OF REPRESENTATION

Barack obama's black presidency has shocked the symbol system of American politics and made the adjective in "representative democracy" mean something quite different than in the past. Obama provoked great hope and fear about what a black presidency might mean to our democracy. His biracial roots and black identity have been a beguiling draw and also a spur to belligerent reaction. White and black folk, and brown and beige ones, too, have had their views of race and politics turned topsy-turvy. What many Americans of all colors believe is that race fundamentally defines America and is a dividing line drawn in blood through the nation's moral map. Many metaphors of race drape the nation's political framework: Barack Obama argued in his famous March 2008 race speech in Philadelphia that slavery is the nation's original sin, and former secretary of state

Condoleezza Rice claimed that racism is our country's "birth defect."[1] Race is the most durable link in the nation's chain of destiny; it is at once a damning indictment of our quest for real democracy and true justice, and also a resilient category of individual and group identity—one that cannot be reduced to either mere pathology or collective pride. Race is both the midwife of glorious achievements like jazz and the black freedom movement and the abortive instrument of Jim Crow and the Ku Klux Klan. Race is the thing we cannot seem to do without—and the thing that we cannot seem to get rid of.

Race is the defining feature of our forty-fourth president's two terms in office. Obama's presidency is a lens to sharpen the details of American ideas about race and democracy. His presidency also raises the question of how much closer the election of a single black man may bring us to a more just and inclusive society. Barack Obama has finally made transparent the idea that our country cannot fully flourish without embracing a black identity that is the quintessential expression of the American character. What we all intuitively sense is that this presidency turns on their ear all the ways we have historically looked at presidencies, and perhaps, even more broadly, at our very democracy. Obama certainly bears what James Baldwin called "the burden of representation."[2] That brilliant phrase refers to the weight and meaning of blackness for individual and collective racial bodies, and for literal and symbolic bodies too. This presidency, unlike all others before it, is analyzed and understood through our obsession with race in the body of the president himself and in the psyche of the nation he governs. A black presidency is undeniably interracial in the same way that Obama's body is composed of black and white genes. Obama's presidency is the symbolic love child of *Notes on the State of Virginia* and *The Fire Next Time.*[3]

Thomas Jefferson and James Baldwin gaze at us from immortal perches separated by two centuries and two races locked in fateful struggle. But Jefferson and Baldwin are separated *only* by time and race; they are united in their unrelenting sexual and political preoccupation with the "other." Jefferson and Baldwin can finally be joined in the full complexity of a conversation about race and American politics across time — a conversation that is constantly evoked but never fully engaged, as if it were held behind doors that are locked to everyone who would participate. And yet Obama is snared in a fascinating paradox: a man seen by many observers as the key to the locked doors of conversation about race is most reluctant to take charge and unlock the treasures of racial insight and wisdom.

What we learn about Obama says a lot about what we learn about ourselves; his racial reality is our racial reality. And it is never, ever static. That truth becomes apparent when we understand just how much we as a nation project our expectations and frustrations onto Obama's presidency, and how he effortlessly represents our deepest doubts and our most resilient hopes. We must concentrate on what Obama says and does — on what speech he gives or what policy he enacts or fails to implement. We must also grapple with what Obama literally means, what his ideas amount to, what veins of ideology or sources of racial imagination he taps when he speaks, and where we travel as a nation by welcoming or resisting the social pathways his presence lays before us.

Obama's presidency represents the paradox of American representation. Obama represents *for* all of us because he stands as the symbol of America to the world. He also represents *to* the American citizenry proof of progress in a nation that has never before embraced a black commander in chief. Yet a third sense of representation has a racial tinge, because Obama is also a representative *of* a black populace that, until his election, had been excluded from the highest reach of political representation. These three meanings of representation are the core of Obama's paradoxical relationship to the citizens of the country he represents: he is at once a representative of the country, a representative of the change the country has endured, and a representative of the people to whom change has been long denied and for whom that change has meant the most.

Of course critics may read "black presidency" as a term that denies Obama the agency and individuality that mark genuine social and moral achievement. To say "black presidency" is already to have reduced Obama's presidency to something less than any other presidency. But the term also imbues the presidency for the first time with the true promise of democracy on which this country was founded. The paradox of representation is thus two-sided: a member of a minority group deliberately excluded from opportunity now stands at the peak of power to represent the nation. The idea of race both qualifies and enhances the representative stature of the presidency. When it comes to race, representation in America is always an internal barometer of privilege, through the exclusion of blacks and others, while at the same time, given how central to our lives race has become, it is also an external barometer of justice.

In *The Black Presidency* I examine Barack Obama's political journey to tell a story about the politics of race in America—our racial limits and possibilities, our tortured past and our complicated present, our moral conflicts and aspirations, our cherished national myths, and our contradictory political behavior. The cultural impact of Obama's lean black presidential frame will be far more enduring than partisan debates about his political career. Obama has changed the presidency itself; the ultimate seat of power has now been occupied for two terms by a man whose body translates in concrete terms our most precious democratic ideals. Obama gives African legs to the Declaration of Independence and a black face to the Constitution. Obama's black presidency cannot be erased by political will even as Congress thwarts his legislation. The paradox of representation Obama symbolizes is not up for judicial review even as the Supreme Court troubles the black vote that helped to sweep him into office.

The existence of a black presidency signals for some people an end to racial categories that have plagued America since 1619. The post-racial urge rises in a society seeking to avoid the pain of overcoming its racist legacy. Obama's presidency has defeated the post-racial myth, not with less blackness but with more of it, though it is the kind of blackness that insinuates and signifies while hiding in plain sight. The presidency is now permanently marked by difference, one that transcends Obama himself and may pave the way for a female president whose gender will be far less noteworthy for Obama's having been the first black president.

A black presidency and the politics of a lived American democracy are like a transmission and its motor: the motor creates the power and the transmission makes the power usable. A black presidency necessarily engages the identity and meaning of an American democracy that was for so long an efficient engine for excluding black participation. Some may worry that the term "black presidency" is code for a delegitimized presidency that undermines democratic institutions and ideas. But Obama's achievement gestures toward what the state had not allowed at the highest level before his emergence: equality of opportunity, fairness in democracy, and justice in society. Our system of government gains more legitimacy when it accommodates demands for justice and adjusts to the requirements of formal equality. Obama's presidency, paradoxically, both critiques and affirms a political order that stymied the ambitions of other black politicians—an order he now heads.

I grapple in *The Black Presidency* with what happens to the psyche and racial identity of a nation when a two-century-old white monopoly on the presidency is broken for two consecutive terms. Then, too, we must ask how and what the blackness of Obama signifies to other blacks. Obama's eight years in office will be referred to as the *only* black presidency until another black person is elected. If the first line in his obituary reads "first [and perhaps only] black president," is he forever fixed in the American mind with a racial reference that he labored hard to overcome? Obama lives with a burden and possibility that no other black person in our history, perhaps in world history, has ever had to shoulder.

A brief survey of other figures might shed light on Obama's unique historical situation. Margaret Thatcher looms large.[4] Thatcher-as-prime-minister is the nearest analogy we have to Obama-as-president. Of course, the biggest difference between Thatcher and

Obama is that Thatcher was overtly ideological and Obama is anti-ideological, the very reason he was electable. There are other differences. Is Thatcher's premiership, these many years later, evaluated as "a woman's leadership" or "the Thatcher Years"? For the first few years of her tenure, not to mention before her election, when she was opposition leader—imagine an American woman in the late seventies as the political and ideological leader of one of our two ruling parties—critics mused about how or whether her gender determined her style of governing. But Thatcher was so hard-line, in truth, heartless, in so many areas—in other words, so stereotypically "masculine"—that in time she was thought of no longer primarily as a woman but as a steely power player, albeit a female one: "the Iron Lady."[5] (Can we imagine a time when Obama would not be seen as black but merely as president? For that matter, should Hillary Clinton or another woman become president, can we imagine her beyond gender on these shores?) Still, by the time she lost power in 1990, British women were not, because of her position, living in a post-gender world, and they still are not today.[6] Yet some of us naïvely believe that Obama's rise has removed race from the national landscape.

Other analogous figures come to mind, including Benjamin Disraeli, the first Jewish prime minister in the United Kingdom, and John F. Kennedy, America's first Catholic president. Each offers instructive similarities. Disraeli's Jewish identity forced him to assure the largely Christian constituency in nineteenth-century Britain that he would not favor Jewish citizens.[7] In the same vein, Obama has not favored blacks, opting, arguably, to underplay their interests in order to reinforce his racial neutrality. Kennedy assured American citizens that he would not take his marching orders from the Vatican.[8] Obama went him one better: he pushed aside the former, if greatly weakened, black political pope, Jesse Jackson, and helped to enshrine a new one, Al Sharpton, while keeping at a distance the Congressional Black Caucus, the archbishops of black politics.

Disraeli and Kennedy had, as did Thatcher, their whiteness, an escape hatch that Obama lacks. If boxer Jack Johnson possessed "unforgivable blackness," then Obama is plagued by inescapable blackness.[9] Disraeli soothed the fears of the masses about his Jewishness, Thatcher toned down her femaleness, and Kennedy downplayed his Catholicism and emphasized instead the catholicity of his politics. All three appealed in their own way to the undergirding whiteness that bound them to their constituencies beyond gender and religious difference. Yet color trumps all for Obama; to have one's presidency examined through the lens of race before any other is as different as Obama's election itself.

Bill Clinton's case is not quite like the other figures', each of whom possesses a quality — ethnicity, gender, religion — that makes their political experiences analogous to Obama's presidency. But the example of Clinton, steeped in the cultural signifiers of blackness rather than race, still offers an intriguing parallel to consider.[10] Toni Morrison and Chris Rock dubbed Clinton the nation's "first black president"; the white politician from Arkansas shrewdly manipulated the meanings and symbols of blackness to his advantage.[11] Clinton strategically embraced blackness to gain the black vote while signaling white suburban voters that he would not bow to Jesse Jackson's leadership.[12] Before his impeachment, Clinton signed a crime bill that sparked a deadly spike in black incarceration and signed into legislation welfare reform that cruelly cut black bodies unable to find living-wage work from public assistance.[13] After his political trial by fire, Clinton embraced Jesse Jackson and played upon black sympathy as smoothly as he blew his sax. Clinton prefigured Obama's even more complicated use of black ideas and black identity while occupying the Oval Office.

Obama, however, stands alone as the only black person to occupy the world's pinnacle of power. What he does, says, and means is as important to the future as it is to our own moment. We must grapple with Obama in the present to set the baseline for his interpretation in the years to come. *The Black Presidency* is my contribution to that goal.

In an Oval Office interview the president granted me for this book, he told me, "In the same way that some of the people who don't like me probably don't like me because of race, there are some people who probably like me because of race and put up with me in ways that they wouldn't if I weren't African American—the folks in African American neighborhoods who identify with me even if they disagree with my policies. And my hope would be that when you wash out those aspects of it, that people are judging me on what I do as opposed to who I am."

The Black Presidency wrestles with the words and actions of a singular human being who rose to the summit of American power; it also measures the racial currents his life captures and conveys, and offers the president informed and principled criticism.

Finally, this book asks, and engages, every complex question suggested by its subject. Is it reasonable to expect more than Obama has offered black people and the American public? What are the salient issues provoked by a black presidency, and how does it affect our ideas of race? How does Obama's relationship to his black elders reflect generational conflicts in fighting for progress in black America? How does Obama's racial identity influence our understanding of his duties? How does the way he speaks reflect the black cultures that molded him? What can we learn from his major race speeches about the ideas that shaped him and the way he confronts racial crises? How does Obama respond to the plague of police brutality that has swept the nation—and the revived racial terror that stalks the land? How does Obama's habit of scolding black America reinforce harmful ideas about black culture? How does Obama's emphasis on law and order, personal responsibility, and respectability politics obscure the structural features of black suffering? What—and who—would it sound like if Obama cut loose and said what he really believes? In *The Black Presidency* I answer these and other questions while confronting Barack Obama's—America's first—black presidency.

HOW TO BE A BLACK PRESIDENT

"I Can't Sound Like Martin"

THE SUNDAY MORNING OF THE MARCH WEEKEND OF EVENTS CEL-
ebrating the fiftieth anniversary of the historic 1965 marches from
Selma to Montgomery, Alabama, was the time in Selma for some
serious preaching. The focus, of course, was on Bloody Sunday, the
fateful pilgrimage that dramatized the violent struggle for the black
franchise and helped push the Voting Rights Act into law less than
six months later. The radiant Sunday was made even brighter by the
presence of so many stars from the black civil rights establishment
who had marched fifty years before. They mingled with present-day
luminaries in the Brown Chapel AME Church, the starting point for
the marches and one of the architectural touchstones in the electrifying film *Selma*. The fact that President Barack Obama was to deliver
what was expected to be a rousing speech on race had been the draw
bringing thousands upon thousands of people to this sleepy southern city still mired in poverty and largely frozen in time.

A few of us sat in the minister's office exulting in the camaraderie and lighthearted banter that black preachers share before the Word is delivered.

"What's up, Doc," the Reverend Al Sharpton, the morning's featured preacher, greeted me.

"What's up, Reverend? Looking forward to your sermon this morning."

I had walked into the church office with the Reverend Jesse Jackson, whose coattails I had much earlier followed into my own ministry and, during his historic run for the presidency, into serious political engagement. I had heard Jackson preach in person for the first time in 1984 on Easter Sunday at Knoxville College in Tennessee. The tall, charismatic leader had cut a dashing figure as he delivered a thrilling sermon-as-campaign-speech in which he criticized President Reagan's military budget, with its priority on missiles and weapons, saying the document represented "a protracted crucifixion" of the poor.[1]

"We need a real war on poverty for the hungry and the hurt and the destitute," Jackson proclaimed. "The poor must have a way out. We must end extended crucifixion, allow the poor to realize a resurrection as well."

Jackson argued that President Reagan had to "bear a heavy share of the responsibility for the worsening" plight of the poor. "It's time to stop weeping and go to the polls and roll the stone away." Jackson also blasted cuts in food stamps, school lunches, and other social programs.

"People want honest and fair leadership," he said. "The poor don't mind suffering," but, the presidential candidate declared, "there must be a sharing of the pain." Jackson clinched the powerful parallel between Christ's crucifixion and the predicament of the poor, especially the twelve thousand folk who had been cut off from assistance, when he cried out that the "nails never stop coming, the hammers never stop beating."

It is easy to forget, in the Age of Obama, just how dominant Jackson had been after Martin Luther King Jr.'s death, how central he had been to black freedom struggles and the amplifying of the voices of the poor.[2] It was in Selma, during the marches in 1965, that a young Jackson was introduced to King by Ralph Abernathy and began to work for the Southern Christian Leadership Conference. He had only later been shoved to the political periphery by the rush of time and the force of events, and viewed as a relic—or worse, as a caustic old man—after he was caught on tape wishing to do away with Obama's private parts. Jackson's weeping visage later flashed on-screen at the celebration in Chicago's Grant Park of Obama's first presidential election. Some viewed Jackson's sobbing as the crocodile tears of an envious forebear. In truth, Jackson was overcome with emotion at a triumph for which he had paved the way. Sharpton was now the nation's most prominent civil rights leader; relations between him and Jackson alternated between frosty and friendly.

Jackson had been Sharpton's mentor as well as mine, and the two embraced in a genial half hug before Sharpton squeezed onto the couch between Jackson and Andrew Young, the former UN ambassador, Atlanta mayor, trusted lieutenant to King—and a father figure of sorts to Jackson. The reunion of Jackson and Young, with Sharpton at the center, was a bit of movement theater. The occasion in Selma had brought together three generations of the bruising patriarchy that black leadership had so often been, with its homegrown authority and blurred lines of succession. I could not let the opportunity pass to quiz Young about his thoughts on Obama and race in the company of his younger compatriots. The elder statesman pitched his views about the president to the home base he knew best: Dr. King and the arm of the movement he had helmed.

"Well, you know, Martin always depended on me to be the conservative voice on our team," Young said, smiling and with a twinkle in his eyes less than a week before his eighty-third birthday. I knew this story, but it was delightful to hear Young regale us with his witty retelling.[3]

"I remember one day Hosea Williams [an aide whom King dubbed his "Castro"] and James Bevel [an aide and radical visionary] were off on their left-wing thing," Young recalled, glancing across at their sometime collaborator Jesse Jackson, who, despite his seventy-three years, had a boyishly mischievous grin etched on his face. "And I was tired of fighting them, so I agreed with what they were proposing." Young gathered himself on the couch, lurched forward slightly, and delivered the punch line with the confidence of a man who had told this story a few thousand times before.

"Martin got really mad at me. He pulled me aside and said, 'Andy, I don't need you agreeing with them. What I need you to do is stake out the conservative position so I can come right down the middle.'" King found it useful to be more moderate than his wild-eyed staff, yet more radical than Young, the designated "Tom" of the group. It might be plausibly argued that Obama's own hunt for a middle ground between Democrats and Republicans was a later echo of some of King's ideological inclinations, a balancing tendency that led historian August Meier to dub the civil rights leader "The Conservative Militant."[4]

I did not quite know what to expect from Young on the topic of Obama; in 2007, when he was a supporter of Hillary Clinton's in the 2008 election, he had pointed to Obama's inexperience and poked fun at his racial authenticity, which he said lagged behind Bill Clinton's blackness. But I suspected the ambassador had come around. It seemed that Young, taking a page from King's book, might travel between Jackson, whose criticism of Obama had been largely subterranean, given his chastened status, and Sharpton, who made a decision never to publicly criticize Obama about a black agenda as a matter of strategy. But Young's brief answer still surprised me for its empathy toward Obama.

"Look, there's a lot on his plate. And he's got to deal with these crazy forces against him from the right. I think that Obama has done the best he could under the circumstances."

Young's answer contained a good deal of wisdom: Obama *has* faced levels of resistance that no president before him has confronted. No president has had his faith and education questioned like Obama. No other president has had his life threatened as much.[5] No other president has dealt with racial politics in Congress to the extent of being denied an automatic raise in the debt ceiling, causing the nation's credit rating to drop. No other president has had a representative shout "You lie!" during a speech to Congress. No other president has been so persistently challenged that he had to produce a birth certificate to settle the question of his citizenship. University of Chicago law professor Geoffrey Stone has argued that "no president in our nation's history has ever been castigated, condemned, mocked, insulted, derided and degraded on a scale even close to the constantly ugly attacks on Obama." Stone says Obama "has been accused of . . . refusing to recite the Pledge of Allegiance, of seeking to confiscate all guns, of lying about just about everything he has ever said, ranging from Benghazi to the Affordable Care Act to immigration, of faking Osama bin Laden's death and funding his campaign with drug money."[6]

Young is right that it has been difficult for Obama to address race and black concerns in such a toxic environment. Obama is wedged between the obstructions of the right and the obsessions of racists who prayed that a black presidency would never come. Some African Americans feared that the obstacles Obama faced would be used as an excuse not to help blacks, lest he appear to pander to his tribe. Obama fretted over striking just the right balance. In a 2007 strategy session at a downtown Washington, D.C., hotel before a Democratic candidates' debate at Howard University, Obama struggled to establish his own voice. "I can't sound like Martin," Obama said. "I can't sound like Jesse."[7] Obama wanted to be himself while acknowledging the storied history that preceded him and made his candidacy possible. The dichotomy is something Obama has struggled with throughout his presidency. And it underlies an even narrower ledge: Obama has searched for the best way to talk about race without raising the ire of whites, but here his struggle has been less acute; he has worried little about losing black support.

Soaring in Selma

On occasion Obama has soared when speaking of race. Those of us who gathered under the surprisingly warm sun to listen to Obama's speech in Selma were galvanized by his words at the foot of the Edmund Pettus Bridge, the scene of the chaotic showdown between peaceful marchers and violent law enforcement. Obama brilliantly summed up black history in the span of a sentence, saying, "So much of our turbulent history—the stain of slavery and anguish of civil war; the yoke of segregation and tyranny of Jim Crow; the death of four little girls in Birmingham; and the dream of a Baptist preacher—all that history met on this bridge."[8] Obama celebrated the "courage of ordinary Americans willing to endure billy clubs and the chastening rod; tear gas and the trampling hoof; men and women who despite the gush of blood and splintered bone would stay true to their North Star and keep marching towards justice."

Obama even found humor in the interracial character of the struggle. "When the trumpet call sounded for more to join, the people came—black and white, young and old, Christian and Jew, waving the American flag and singing the same anthems full of faith and hope. A white newsman, Bill Plante, who covered the marches then and who is with us here today, quipped that the growing number of white people lowered the quality of the singing." Peals of laughter mixed with applause. Obama insisted, too, that these men and women were patriots. Perhaps he was thinking of the hateful charge of disloyalty to his own country he had endured, the claim that he failed to be truly or fully American weighing on his chest as he inhaled the crisp Selma air and exhaled a ringing affirmation of the love of nation displayed by those folk on a bloody bridge fifty years before. "What greater expression of faith in the American experiment than this, what greater form of patriotism is there than the belief that America is not yet finished, that we are strong enough to be self-critical, that each successive generation can look upon our

ecide that it is in our power to remake this nation
n with our highest ideals?"

the criticism of country as an act of love, one
merican patriots long ago, an act every bit as
elfare of a nation that "is a constant work in
s of those whose ardent support is more eas-
g this country requires more than singing its
ncomfortable truths," he said. "It requires the
, the willingness to speak out for what is right,
s quo. That's America."

fter Obama encouraged patriotic disruption, I
t veteran Bernard Lafayette leave his VIP seat
ckson and attend to a group of young black
on, Missouri, the self-titled Lost Voices Group,
who were stationed outside our roped-off area and who sought to
shake up the status quo and interrupt the president's speech. The
activists were beating a drum rhythmically and chanting "We want
change, we want change" and "Black Lives Matter," although they
weren't close enough to the stage to compete with the president's
richly amplified voice. One of the protesters complained to Lafayette that no one would listen to their demands; the elder statesman, in turn, empathized with the young people and told them that
he had been arrested twenty-seven times during sixties protests, and
had even been the target of an assassination plot during that time
in Selma. After the activists fell silent, Obama's words continued to
echo through the adoring crowd.[9]

The president linked his own success to the hard road traveled by
the dusky patriots he had exhorted, noting how "the change these
men and women wrought is visible here today in the presence of
African Americans who run boardrooms, who sit on the bench,
who serve in elected office from small towns to big cities; from the
Congressional Black Caucus all the way to the Oval Office." Obama
brought full circle his initial appearance in Selma in 2007, when he

was first running for the Oval Office, and the fulfillment of that promise in his presidency. Obama had come to Selma then as a candidate and defiantly refuted those who suggested that his African heritage kept him from sharing the civil rights legacy: "So don't tell me I don't have a claim on Selma, Alabama. Don't tell me I'm not coming home when I come to Selma, Alabama. I'm here because somebody marched for our freedom. I'm here because y'all sacrificed for me. I stand on the shoulders of giants."[10]

Obama acknowledged in his second Selma speech the contemporary plagues that tarnished the legacy of that city—especially the rash of police brutality and killings of unarmed black folk. The president said that two temptations beckon us. The first is to say, on the one hand, that since Selma and the sixties nothing has changed. Obama acknowledged that the injustice of Ferguson, Missouri—where uprisings followed the death of an unarmed black youth, Michael Brown, at the hand of a white policeman, and where the Department of Justice found the city's police department plagued by racism—was real, but that while it "may not be unique," it is "no longer endemic. It's no longer sanctioned by law or by custom. And before the Civil Rights Movement, it most surely was." Of course, many would argue that deeply entrenched bias still reflects informal habits that have yet to be uprooted, even if they are not officially sanctioned, and that the repeated offenses of law enforcement against blacks and other minorities entail a violent bigotry that infects the systems of policing in our nation. On the other hand, Obama said, it is wrong to assume that an episode like Ferguson is "an isolated incident; that racism is banished; that the work that drew men and women to Selma is now complete."

Obama condemned efforts to weaken the Voting Rights Act that was Selma's gift to black folk and the nation. "Right now, in 2015, fifty years after Selma, there are laws across this country designed to make it harder for people to vote. As we speak, more of such laws are being processed. Meanwhile, the Voting Rights Act, the culmination of so much blood, so much sweat and tears, the product of so much sacrifice in the face of wanton violence, the Voting Rights Act stands weakened, its future subject to political rancor." Obama also praised young people in attendance and across the country: "You are America. Unconstrained by habit and convention. Unencumbered by what is, because you're ready to seize what ought to be."

Trayvon Martin's Trial, and Ours

It was easily Obama's best presidential speech on race at the time because of its eloquence, and, sadly, because there were few others to compare it to, even though he had by then been in office for six years. Obama often has been loath to lift his voice on race lest he be relegated to a black box, although his reluctance has kept the nation from his wisdom and starved black folk of the most visible interpreter of their story and plight, an interpreter who also carries the greatest political clout in the nation's history. His radio silence often sent the wrong signal that race, and black concerns, did not count as much as other national priorities. A single event in 2012 smoked Obama out of his presidential cubbyhole of racial non-engagement and thrust him into his bully pulpit to, in part, define, and then defend, black people—really, to represent them, an extraordinary feat in itself. It was the epic grief that gripped black America with the not-guilty verdict in 2013 in the trial of self-styled neighborhood watchman George Zimmerman in Sanford, Florida, for the fatal shooting of black teen Trayvon Martin. That verdict, and the persistent injustice it highlighted, contrasted sharply with the narrative equating Obama's ascent with the end of race in America.

Obama spoke about Trayvon Martin to explain to white Americans why so many black folk were enraged over the verdict. Many white conservatives viewed Obama's "one-sided" explanation of black suffering—a radical departure from the tough blows he had thrown black people's way in most of his public pronouncements on blackness—as a surly betrayal of his racial agreement.[11] Some whites believed that agreement was: do not speak much on race, and when you do, go after your own, and offer the blandest platitudes possible about the progress made and the racial work that remains to be done.[12]

The most visible case of Obama being challenged to speak up for our most vulnerable black citizens—as a president should do for all citizens who suffer—occurred when he marched purposefully into the White House pressroom and offered an off-the-cuff, from-the-heart oration to the nation in the aftermath of the not-guilty verdict for George Zimmerman in July 2013. Until his 2015 Selma speech, it was Obama's most extensive treatment of race since coming to office. Yet it was a speech made only after Obama's written statement a few days earlier fell woefully short in calling for calm observance of the law in protests that greeted the verdict.[13] The mounting disappointment finally squeezed him into interpreting black pain for the country's benefit.

Obama acknowledged that the verdict, a staple of our justice system, had been rendered, but still, he wanted to supply context to the outrage millions of black people felt at the jury's decision. The president reminded the nation that when he first spoke about the killing of Trayvon Martin in March 2012, he'd said Trayvon could have been his son. Obama widened the reach of racial intimacy and got even more personal this time: "Trayvon Martin could have been me thirty-five years ago."[14] The importance of Obama's identifying with a young man whose memory had been soiled in cyberspace as a wanton thug who'd gotten what he had coming to him could not be overstated. Obama cast his fate with Trayvon's and thus threw the enormous weight of his office behind the black teen from Florida. Trayvon Martin could have been profiled and killed at any point in our nation's history, given our long bout with bigotry. But this youth's death at this time meant something to a man who had been profiled since he'd won the presidency, a man who now looked at what might have been his younger self lying prostrate on the ground in Florida, until the difference between them disappeared.

Obama spoke of the micro-aggressions that eat at black folk—being followed in the department store while shopping, people locking their car doors when you approach, or women clutching their purses when you get on the elevator. Those experiences colored how blacks interpreted the verdict and the criminal justice system. "The African American community is also knowledgeable that there is a history of racial disparities in the application of our criminal laws—everything from the death penalty to enforcement of our drug laws. And that ends up having an impact in terms of how people interpret the case."

After saying that black people were aware of how young black men are "disproportionately both victims and perpetrators of violence," Obama argued that historical context is critical to understanding the violence they commit and endure—especially the poverty and dysfunction that shape their lives but which are often unrecognized by the larger society. Obama also expressed the frustration of seeing black youth "painted with a broad brush" as violent creatures to justify treating them differently. He summed up the injustice of their different treatment by drawing a contrast: "And that all contributes I think to a sense that if a white male teen was involved in the same kind of scenario, that, from top to bottom, both the outcome and the aftermath might have been different."

Obama posed the question of where the nation might go beyond vigils and protests. Besides a review of the case by the Department of Justice—which later found that it could not prove Zimmerman had intentionally violated Martin's civil rights—he pointed to a few things that we as a nation might do to work through the pain of our racial fracture. First, he suggested efforts to "reduce the kind of mistrust in the system that sometimes currently exists," a theme Obama has repeated after nearly every deadly interaction between the cops and black people over the last few years. He also cited "potential racial bias" in law enforcement in ways he has not always made clear. Second, the president argued for an examination of state and local laws with an eye to whether "they may encourage the kinds of altercations and tragedies that we saw in the Florida case, rather than defuse potential altercations." Obama proposed that we find alternatives to the "stand your ground" law, alternatives that would counsel exit from a potentially dangerous situation. A national effort was mounted to overturn "stand your ground" laws after Martin's death—although an appeal to that law was not made by the defense in the Zimmerman case. For those who resisted the notion that these laws should be reviewed, Obama offered a point made by many black critics. "If Trayvon Martin was of age and armed, could he have stood his ground on that sidewalk?" Obama poignantly and calmly queried. "And do we actually think that he would have been justified in shooting Mr. Zimmerman who had followed him in a car because he felt threatened? And if the answer to that question is at least ambiguous, then it seems to me that we might want to examine those kinds of laws."

Third, Obama offered the kernel of his "My Brother's Keeper" initiative,[15] saying, "We need to spend some time in thinking about how do we bolster and reinforce our African American boys." Obama argued against "some grand, new federal program." He encouraged the nation instead to "do some soul-searching" in the wake of the verdict. "I haven't seen [it] be particularly productive when politicians try to organize conversations. They end up being stilted and politicized, and folks are locked into the positions they already have." Instead Obama wished that Americans would be more honest in their own homes, churches, and workplaces, asking themselves a couple of questions: "Am I wringing as much bias out of myself as I can? Am I judging people as much as I can, based on not the color of their skin, but the content of their character?"

Perhaps Obama could have sparked people to be more honest about their personal lives if he had spoken more forthrightly about the complicated jigsaw puzzle that is race and what we as Americans might do to make things fit more justly. Obama seems to believe, despite not buying the post-racial myth or the delusion that racism is dead, that "each successive generation seems to be making progress in changing attitudes when it comes to race." It's a belief that, while noble, may not be borne out by the facts: a recent study proves that, with the exception of interracial dating and marriage, "millennials are just as prejudiced as their parents."[16] Obama's confidence that "kids these days, I think, have more sense than we did back then, and certainly more than our parents did or our grandparents did" simply is not supported by the facts and can't be the basis of honest racial conversation.

The Audacity of Nope

Obama's skepticism about how much he could positively impact race as president emerged when he sparked a controversy by saying that the Cambridge, Massachusetts, police "acted stupidly" in arresting black Harvard professor Henry Louis Gates Jr. for disorderly conduct. Gates had returned home from a trip to China to discover that his door was jammed, and as his driver helped him gain entry to his house, a passerby called the police, thinking it might be a break-in. The ensuing conflict between Gates and the police officer who responded to the call led to his arrest and a national debate about race and law enforcement. I address the Gates affair more extensively later on, but for now I will note that the Gates confrontation made Obama very hesitant to speak on race and led him to three fateful conclusions. One, never speak of race in a way that holds whites even partially responsible for black suffering. The subject of white guilt of any sort—even in circumstances of clear white culpability—is to be avoided at all costs. This is another way of saying that race is primarily the business and burden of blacks. Two, although they read it quite differently, both white and black communities are eager for Obama to excoriate perceived black error—for instance, in his warning to Morehouse College graduates against using racism as an excuse for failure—and to damn black pathology, such as absentee fathers. In black life such gestures are often read as tough love; in white America they are seen as heroic battles against black deficiency. Third, when the structural features of black suffering cannot be ignored, it is best to soften the blow with ample mentions of black criminality or black moral failure.

When Obama addressed for the second time the racially charged situation in Ferguson, he held true to this racial template.[17] He did not mention that a white policeman was responsible for the death of unarmed black youth Michael Brown, the crux of the uprising in Ferguson, although he mentioned people of color several times, reinforcing the view that race is a black issue, a black obligation, and not at all a white responsibility. Obama tiptoed around a discussion of structural forces and unjust practices—such as a racially charged criminal justice system, or the fact that black kids are expelled from school at higher rates than whites. He took on the issue in the politest terms possible, almost parenthetically, fearing to offend police forces throughout the nation, which are responsible for ungodly numbers of dead black bodies. True to form, Obama shifted from a discussion of the police to the subject of black responsibility.

If Obama's delivery on race is sad and disappointing, the failure of most black Americans to hold him accountable is no less so. Blacks are understandably reluctant to criticize the president. It is easy to see why. The election of Barack Obama in 2008 flooded black America with euphoria. There had been forty-three American presidencies stretching back to 1789, and not a single one was black. Two hundred and twenty years after George Washington was sworn in as the nation's first president, Barack Obama was sworn in as the nation's first black president. Obama's win forced syntax and symbolism to get hitched in a shotgun wedding: the adjective "black" may have modified the noun "president," but the change went far beyond a single man's life or office and promised to transform the meaning of race in America.

Black America beamed with pride when Obama became the most powerful black man in history. Many black folk believed that Obama's victory was their triumph. They desperately hoped that his election was a gesture of respect for the black masses. Their hopes persisted even when mainstream media and political observers hailed Obama's win as a beacon of post-racial America. Most black people were skeptical and knew that one black man moving into the nation's most spectacular public housing could not defeat racism.

Obama's historic feat did not put a smile on every citizen's face. Republican leadership did not want history to repeat itself and prayed for a single-term black presidency.[18] The far right quickly got sick of Obama and took in plenty of tea parties as a cure. The political fringe tried to spook the nation into fearing Obama's dangerous otherness, caricaturing him as a chimpanzee and, in a sobering denial of his humanity, as a black box, a dark cipher. Obama could not persuade his opponents that darkening the corridors of power posed no threat to the American way of life or the fortunes of white civilization. Hitting the links with Republican politicians for a friendly game of golf did not stop his critics from routinely teeing off on him or slicing his policies. At the same time, Obama's power establishment comfort zone angered lots of progressives, who believed that he gave cover to American empire. Obama's race seemed to matter most when it should not have mattered at all, especially in a society that had supposedly moved beyond color with his election. In such an atmosphere most blacks were hesitant to say a discouraging word about Obama.

The failure of most black Americans to insist that Obama address race as a matter of course, honestly, accurately, has done two bad things. It offered Obama the preferential treatment he did not want, and it guaranteed that black Americans would continue to get what they did not deserve: a president who was afraid to address their concerns effectively. When he was campaigning for president, Obama was asked why Martin Luther King Jr. would have endorsed him. "Dr. King would [not] endorse any of us," he answered. Instead, Obama argued, King "would call upon the American people . . . to hold us accountable." Obama concluded that "change does not happen from the top down; it happens from the bottom up," with activists "arguing, mobilizing, agitating and ultimately *forcing* elected officials to be accountable."[19] Obama, like all presidents, addressed race only when faced with principled critique and persistent pressure.

Obama's views on race and America have drawn considerable criticism from scattered quarters of black culture. How should black leaders and thinkers behave toward a president who is adored by the black masses and loathed by large numbers of whites? Pretending that it makes no difference that Obama is the first black president is either naïve or foolish. It is like ignoring the bruising battles that Jack Johnson endured to become the first black heavyweight champion and the grievous price he paid for being good enough to win. It is to ignore how Jackie Robinson was spiked with bigotry in baseball, or the difficulties endured by hundreds of other black firsts, whether they were medical pioneers or magazine editors. Black pride in Obama clashed with white resentment of his rise to power. The black urge to protect Obama is fueled by the belief that blacks must do everything possible to shield the president from racist rhetoric while refraining from criticism that might increase his vulnerability.

Black protectionism presents black leaders, and black critics, too, with a tough choice: either side with an unfairly besieged president in a show of black loyalty or risk retaliation by black folk who hear a whisper of dissent as a hurricane of hate. Congressional Black Caucus chair Emanuel Cleaver, when asked in 2012 if Obama was being spared criticism for high black unemployment rates because of his color, explained the reluctance to go after the president: "Well, I'm supposed to say he doesn't get a pass, but I'm not going to say that. Look, as the chair of the Black Caucus I've got to tell you, we are always hesitant to criticize the president. With 14 percent [black] unemployment if we had a white president we'd be marching around the White House." Cleaver suggested that ethnic identification and racial solidarity would prompt most groups to avoid attacking the president, as blacks have done with Obama. "I . . . don't think the Irish would do that to the first Irish president or Jews would do that to the first Jewish president . . . we're human and we have a sense of pride about the president." Cleaver admitted that Obama exploits black pride too. "The president knows we are going to act in deference to him in a way we wouldn't to someone white."[20]

Black leaders and thinkers are often caught between white revulsion and black reverence, making it difficult to offer the president sensible pushback and enlightened criticism. Besides black conservatives, who largely echo the views of their white counterparts, at least five groups of blacks have criticized Obama: members of Congress who support the president but occasionally press him for more resources for their constituents; progressive intellectuals and ministers who call on history and politics to chastise the president for his neglect of black interests and suffering; feminist critics and public intellectuals who seek to combat the exclusion of women and girls from national and presidential conversations on race; critics on the far left who lambaste Obama as a prop for American imperialism abroad and a puppet of domestic financial exploitation; and public intellectuals and activists who often differ with the president yet offer either principled support or ad hominem attacks.[21]

You Were Always on My Mind

Obama's nearly unanimous black support could not completely muffle grumblings about the ebony wunderkind among some black leaders. Senator Obama not only kept his Congressional Black Caucus colleagues at arm's length but also continued the practice once he became president. Although black adoration has kept a lid on black disgruntlement with the president, Black Caucus members have periodically questioned Obama's commitment to the black base that swept him into office.

A few months after Obama's 2012 reelection victory, Congressional Black Caucus chair Marcia Fudge sent a letter to the president chiding him for the lack of diversity in his second-term cabinet choices. At the time of Fudge's letter, it had been at least 675 days since the president had met with his former Black Caucus colleagues.[22] At the start of his second term Obama appointed nine new cabinet members, including three women and one Latino. Women's groups had complained of low numbers among his cabinet members and were promptly rewarded.

"I am concerned that you have moved forward with new cabinet appointments and yet, to date, none of them have been African American," Fudge wrote in her letter to Obama.[23] "You have publicly expressed your commitment to retaining diversity within your cabinet. However, the people you have chosen to appoint in this new term have not been reflective of this country's diversity."

Fudge noted that CBC members had received many complaints from constituents questioning why Obama had failed to name blacks to his cabinet. "Their ire is compounded by the overwhelming support you've received from the African American community." When the CBC was finally granted a meeting with the president in July 2013, their first since 2011, Fudge had reason to change her tune. "What I did today is I thanked the president for [Transportation Secretary] Anthony Foxx and for [Federal Housing Finance Agency director] Mel Watt," Fudge said, suggesting that her tough public criticism had worked.[24]

The CBC's dilemma was perfectly summed up during its 2011 jobs tour, when California congresswoman Maxine Waters pleaded with black voters at a town hall meeting in Detroit to free black Democratic politicians to criticize Obama for his failure to help the black community. Obama was then on a bus tour of Midwest rural cities ravaged by deindustrialization; the president had no black communities on his itinerary. "We don't put pressure on the president because y'all love the president," Waters admitted. "If we go after the president too hard, you're going after us." Waters laid out her views in a fiery oration:

> When you tell us it's alright and you unleash us and you tell us you're ready for us to have this conversation, we're ready to have the conversation. The Congressional Black Caucus loves the presi-

dent too. We're supportive of the president but we're getting tired ya'll . . . [W]e want to give the president every opportunity to show what he can do and what he's prepared to lead on . . . but our people are hurting. The unemployment is unconscionable. We don't know what the strategy is. We don't know why on this [Midwest] trip that he's [on] in the United States now, he's not in any black community . . . we don't know that.[25]

Congressman Cleaver supported his colleague. "It's not personal," Cleaver insisted. Waters and other CBC members were "attacking his policies, or lack thereof, with regard to this gigantic unemployment problem among African-Americans. If we can't criticize a black president, then it's all over."[26]

Black political leaders like Marcia Fudge, Emanuel Cleaver, and Maxine Waters have offered criticism of Obama while remaining respectful of his position and conscious of the wishes of their black constituencies. Many black politicians, however, are motivated to protect Obama because the white backlash against him is inherently unjust. In protecting the president, they are also protecting the race, black power, and, indirectly, themselves: black politicians increase their standing with black constituencies that love Obama and wish to protect the legitimate black exercise of power in the White House. If black politicians believed that Obama might return the favor of support, they have, for the most part, been sadly mistaken.

Obama has from the start had a vexed and ambivalent relationship with his black political elders. Despite his community organizing efforts, the Ivy League graduate, at the outset, was viewed suspiciously in black circles as an interloper who got his comeuppance through a drubbing at the polls by Bobby Rush, the seasoned former revolutionary and later Chicago congressman whose Illinois First Congressional District seat the young Obama brazenly sought in 2000. Once Obama had the presidency fixed in his sights, he hushed black doubts by boldly laying claim to bloodstained freedom struggles in Selma and Memphis. He quickly surged on a wave of black love past the sinking fortunes of Hillary Clinton. Along the way Obama rubbed shoulders with and ruffled the feathers of the black political elite, enduring sour grapes, raised hackles, arched eyebrows, standoffish vibes, and wait-and-see attitudes. Obama brashly took his case straight to the masses and forced even Clinton loyalists like John Lewis to switch sides and support the new kid on the block.

Obama upped the ante by stealing scenes from his elders' dramatic history and using their lines to benefit his play for power, casting himself as Joshua, Moses's successor in the Bible, aiming to contrast an older generation's fiery imperatives to his generation's cooler delivery. Obama's clever distinction nonetheless carried risks: Joshua was charged with duties that tied him more closely to his elders than Obama might be willing to admit. Some black figures were willing to press Obama on his failure to embrace the bond between the two generations.

In 2013 Kevin Johnson, a prominent Philadelphia pastor and self-professed "strong supporter of the president" who worked "hard to get him elected in 2008 and 2012," penned a controversial and widely discussed opinion piece, "A President for Everyone, Except Black People," for the local black newspaper, the *Philadelphia Tribune*.[27] Johnson echoed Marcia Fudge's complaint that Obama's second-term cabinet lacked black leaders and advisers. Johnson argued that "when one compared the first African-American president to his recent predecessors, the number of African-Americans in senior cabinet positions is very disappointing: Clinton (7); Bush (4); and Obama (1). Obama has not moved African-American leadership forward, but backwards." Johnson admitted that while gaining senior cabinet positions does not guarantee economic advance for black folk, it at least puts them in the driver's seat, though with "President Obama, we are not in the driver's seat—or in the car." Johnson argued that such a state of affairs was "disrespectful" to the black community that voted "96 percent for President Obama in 2008 and 93 percent in 2012."

Johnson contended that an objective survey of Obama's presidency proved that blacks "are in a worse position than they were before [Obama] became president. At the end of January 2009, un-employment for African-Americans was 12.7 percent," while four years later "the situation is worse, and unemployment is higher at 13.8 percent." Johnson said he'd supported Obama not because he was black, but because he was the best person to "empathize, un-derstand, and develop policies to help the African-American com-munity, the poor, and previously under-represented communities." A disappointed Johnson argued that Obama has not simply "failed the Black community," but his agenda "appears to be for everyone except Black people, his most loyal constituency." He concluded that if Obama "does not make some changes soon, at the end of his presidency he will be known as a historical leader—the first African-American president, but not a transformational leader—the presi-dent who truly uplifted and catapulted Black people from cycles of poverty, unemployment, illiteracy, and despair."

Johnson's commentary got him into hot water. He had been in-vited to deliver the baccalaureate sermon at his alma mater, More-house College, the day before Obama was slated to give the college's 2013 commencement address. But after Johnson's opinion piece ap-peared, Morehouse president John Wilson, who previously headed the White House Initiative on Historically Black Colleges and Uni-versities, informed Johnson that his address would be changed to a multi-speaker event in order to air a broader set of views. Johnson declined to participate under the new format and charged in a letter to Wilson that the college president had "disinvited" him because of his tough talk on Obama. After several alumni of the all-male black college criticized Wilson for his decision, Johnson was reinstated and gave his sermon.

A year before Johnson's article, an opinion piece in the *Washing-ton Post* by Columbia University professor Fredrick Harris had also stirred quite a response with a provocative title: "Still Waiting for Our First Black President." Harris argued in the essay, and in the book from which it was adapted, *The Price of the Ticket: Barack Obama and the Rise and Decline of Black Politics*,[28] that when he was trailing in the race against Hillary Clinton, Obama discussed issues like "racial injustice in front of black audiences" and supported "targeted and universal policies to address racial inequality." Harris claimed that this is "the Obama who has been forgotten, who all but disappeared later in his 2008 campaign and during his presidency." He contended that Obama "has pursued a racially defused electoral and governing strategy, keeping issues of specific interest to African Americans—such as disparities in the criminal justice system; the disproportionate impact of the foreclosure crisis on communities of color; black unemployment; and the persistence of HIV/AIDS—off the national agenda." Harris charged that, far "from giving black America greater influence in U.S. politics, Obama's ascent to the White House has signaled the decline of a politics aimed at challenging racial inequality head-on." He concluded that Obama "may be our first gay president"—referring to a 2012 *Newsweek* magazine cover acknowledging Obama's strong support for gay folk—"but if a focus on racial inequality matters at all, we're still waiting for our first black one."[29]

If Harris was waiting for the first black president, some black feminist critics and public intellectuals argued that at least black men got presidential support with Obama's "My Brother's Keeper" initiative. The president crafted the program to improve the fate of young boys and men of color after the violent, untimely deaths of Trayvon Martin and other young black males. Journalist and public intellectual Brittney Cooper—who also criticized Obama for his brand of respectability politics, and for his paradoxical version of American exceptionalism[30]—was at first ambivalent about My Brother's Keeper.[31] Cooper acknowledged that black boys stand in need of help because they suffer high rates of school suspension, incarceration, and unemployment—and low rates of high school and college graduation—and commit a large percentage of violent crimes. She noted that it is black women "doing the fighting, the organizing, the praying, the rearing, the fussing, the protecting, the loving. And the losing. Black women have been their brothers' biggest and best keepers."[32] Cooper wished that "black men—Barack Obama in-cluded—had the kind of social analysis that saw our struggles as deeply intertwined." She highlights numbers that show just how dire the situation is for black girls:

> According to the African American Policy Forum, black girls are suspended at a higher rate than all other girls and white and La-tino boys. Sixty-seven percent of black girls reported feelings of sadness or hopelessness for more than two weeks straight com-pared to 31 percent of white girls and 40 percent of Latinas. Single black women have the lowest net wealth of any group, with re-search showing a median wealth of $100. Single black men by con-trast have an average net wealth of $7,900 and single white women have an average net wealth of $41,500. Fifty-five percent of black women (and black men) have never been married, compared to 34 percent for white women.[33]

Cooper praises Obama's efforts to secure his post-presidential legacy by forging policies—beyond the Affordable Care Act—that assist the demographic groups most responsible for landing him in the Oval Office. These include LGBTQ people (a 2014 executive or-der that prevents federal contractors from discriminating on grounds of sexual orientation or gender identity), men of color (My Brother's Keeper, broadened clemency for low-level drug offenders, a hike in the federal minimum wage), and Latino families (the 2014 pledge to review Obama administration deportation policies that wreaked havoc in Latino communities). But Cooper justly chafes at Obama's neglect of black women and girls, his greatest voting demographic.[34]

Arguing that black women "have been the subject of no execu-tive orders, no White House initiatives and no pieces of progressive legislation," Cooper writes:

> Ninety-six percent of black women voters voted for Obama com-pared to 87 percent of black men. Seventy-six percent of Latinas voted for the president compared to 65 percent of Latino men.

Though black and other women of color who are a part of the
LGBTQ community will benefit from [the 2014] executive order,
no initiative has explicitly addressed the structural issues of racism, classism, poor education, heavy policing and sexual and domestic violence that disproportionately affect black and Latina
women. As a black woman who voted twice for this president,
despite some misgivings, I find myself wondering how we will fit
into the legacy of progressive policy initiatives that the president
is trying to craft as part of his exit strategy.[35]

Kimberlé Williams Crenshaw, Columbia University law professor
and executive director of the African American Policy Forum, whose
study on black women and girls is the source that Cooper cited, offered a bracing opinion piece in the *New York Times* whose title perfectly captured her argument: "The Girls Obama Forgot." Crenshaw
wonders if "the exclusion of women and girls is the price to be paid
for any race-focused initiative in this era." Addressing the needs of
black men, she argues, "hits a political sweet spot among populations
that both love and fear them," while "erasure of females of color is
regarded as neither politically nor morally significant." Crenshaw
acknowledges that ignoring black women and girls is nothing new;
it grows from the belief that black men are at exceptional risk to
the plagues of racism and the unfounded notion that black women
are faring much better by rough comparison, one that, Crenshaw
maintains, is supported by evidence that "is often illogical, selective
or just plain wrong." She notes that "like their male counterparts,
black and Hispanic girls are at or near the bottom level of reading
and math scores . . . They also face gender-specific risks: They are
more likely than other girls to be victims of domestic violence and
sex trafficking, more likely to be involved in the child welfare and
juvenile justice systems, and more likely to die violently. The disparities among girls of different races are sometimes even greater than
among boys."[36]
Crenshaw questions the way some observers use the high rates of

black male incarceration to justify the exclusive focus on black men and boys: "Is their point that females of color must pull even with males in a race to the bottom before they deserve interventions on their behalf?" Crenshaw also underscores the paradox in the failure of the My Brother's Keeper initiative to gather information about women and girls of color, which feeds the misperception that they are relatively better off than their male counterparts: "The exclusion of girls of color from the data collection means that there will be fewer 'evidence based' interventions for girls—because there was no interest in marshaling evidence to support interventions for them in the first place." She argues that supporters of My Brother's Keeper use the familiar analogy of the canary in the coal mine to scrutinize the experiences of men and boys of color while leaving aside women and girls: "But the point of the canary's distress was to alert everyone to get out of the mine, not to attend to the canary and ignore the miners. Implicit in rescuing only the males is the idea that the mine itself isn't the problem—and that females are resilient enough to survive the toxic air or can hold their breath and wait. What needs to be fixed are not boys per se, but the conditions in which marginalized communities of color must live."[37]

If Cooper's and Crenshaw's progressive perspectives, like those of Kevin Johnson and Fredrick Harris, aimed to push Obama in the right direction, the left-wing critics from the *Black Agenda Report* have consistently lambasted Obama as the genial face of American empire. In "2007: The Year of Black 'Media Leaders'—Especially Obama," *Report* executive editor Glen Ford argued that "Barack Obama's corporate-made-and-financed presidential campaign is the product of three distinct factors."[38] First, there were corporate decisions made a decade earlier to provide media and financial support to "pliant Black Democrats that can be trusted to carry Wall Street's water." Second, there was the desire of whites to prove they were not racist—and to dispute black claims of persistent racism—by voting for a black man. Third was the willingness of black folk to ignore the political stances of candidates in order to embrace the philosophy of black faces in high places. "A President Obama would, of course, be the zenith of such narrow, non-substantive, objectively self-defeating visions," wrote Ford. In 2012, *Report* managing editor Bruce Dixon, in "Tired Old So-Called Leftists Give Same Old Excuses for Support-ing Obama in 2012," argued that black America formed a "veritable wall around the First Black President," but that this wall does not protect Obama from "racists or Wall Street predators or Pentagon warmongers." When black Americans "stifle our own tongues and circle the wagons trying to silence critics of the White House we only protect the president and his party from accountability to their supposed base: us."[39]

In the same year Ford accused the Obamas of being "a global capi-tal–loving couple, two cynical lawyers on hire to the wealthiest and the ghastliest," who are "no nicer or nastier than the Romneys and the Ryans, although the man of the house bombs babies and keeps a kill list."[40] Ford also blasted leftist icon Angela Davis for declaring that Obama's triumph was a "victory, not of an individual, but of . . . people who refused to believe that it was impossible to elect a person, a Black person, who identified with the Black radical tradition." Ford charged that Davis had "soiled herself, and done a terrible injustice to black history and tradition," and that "she has diminished herself and insulted our people for the sake of a president who doesn't give a damn for their history or their future."[41] The *Report* also argued that Fredrick Harris's criticism of Obama is insufficiently leftist because his book fails to engage "the plutocratic political cash and the many sided corporate-imperial establishment—to the unelected and inter-related dictatorships of money and empire—that pre-select 'viable' presidential candidates for popular 'choice' in the first place."[42]

It is reasonable for a black leftist to challenge Obama's rightward drift into what public intellectual Cornel West terms "a Wall Street presidency, a drone presidency, a national security presidency," in which crooked executives and torturers go unpunished.[43] But vitu-peration clouds West's political stances no matter their insight or virtue. West has conducted campaigns of vitriol against Obama in the name of social prophecy. White leftists and liberals are angry with Obama too, but figures like Michael Moore, Roger Hodge, and Diane McWhorter have not resorted to the epithets that stain West's analysis.[44] West skewers Obama as "a black Mascot of Wall Street oligarchs," a "black puppet of corporate plutocrats," and a "Rockefeller Republican in blackface."[45] He has also accused other intellectuals and leaders of selling their souls to Obama, especially the Reverend Al Sharpton.

West has nastily accused Sharpton of "prostituting" himself for the president, and for being "*the* bonafide House Negro of the Obama Plantation," a prospect that would have Martin Luther King Jr. "turning over in his grave."[46] Sharpton told me in an interview that, on the contrary, he is operating in the venerable tradition established by an earlier generation of leaders: "My relationship with Obama is like Dr. King's was with Kennedy. They were close enough for Kennedy to warn him about the FBI's surveillance of figures in the movement." Sharpton chuckled. "Hell, Obama ain't told me to watch out for nothing! So I'm not even as close to Obama as King was to Kennedy. But I have the freedom to do whatever I want to do, including criticize him."

Sharpton has been taken to task for allegedly saying on the television newsmagazine show *60 Minutes* that he would never publicly criticize Obama. "I never said that I would never criticize Obama," Sharpton told me; "[*60 Minutes* correspondent] Lesley Stahl said I said it."[47] Instead, Sharpton contends, "I said I would never criticize him for not having a black agenda." Sharpton believes it his duty as a civil rights leader to pressure the president to embrace racial justice, and that Obama's role is to govern with sensitivity toward an agenda brought to him by black leaders. Sharpton says that he and Obama have differences but that he doesn't exhibit meanness in expressing criticism: "I disagree with [the president] on Afghanistan. And . . . I don't understand why he couldn't immediately close Guantanamo

Bay. He explained in a meeting the military reasons, [but] I don't see any reason not to do it. So we've had debates — friendly, but firm. And we agree to disagree."

As for Obama's policies on race, Sharpton notes: "I always felt in the first term that they were too cautious. They got a little better in the second term." Sharpton maintains that Obama is not being disingenuous when he is highly optimistic about racial progress — though critics view it as profound political naïveté. "He really does think everything is together," says Sharpton. "But I think he underestimated the venom and rage that were coming at him. And I think as time went on, he began to understand [that] we weren't imagining that stuff; there is still real [racism] in this country."

West and his former radio cohost Tavis Smiley also criticized Sharpton for not demanding that the Department of Justice immediately charge George Zimmerman with a civil rights violation in connection with the shooting of Trayvon Martin. "The difference between me and others who commentate is I'm representing real people," says Sharpton. "I've got a real [civil rights] membership. They have nothing. So they're just shooting off at the mouth. I've got to worry whether [criticizing the president and the attorney general] is more important than making sure that Trayvon Martin's family gets a meeting with the Justice Department." Sharpton also argues that Obama has not actively opposed black interests in the way that Bill Clinton did when he signed welfare reform and a crime bill that leveled black America. Still, black America gave Clinton a pass because he offered compensating virtues in making cabinet appointments and preserving affirmative action; Sharpton contends that Obama is surely no less worthy of the black benefit of the doubt.

Sharpton admits that he is in an unprecedented position: for the first time, the nation's premier civil rights leader must coexist with a black commander in chief who is extremely popular with the black masses: "I think we developed a respect for each other's role. We understand we are not the same thing . . . But I think he has grown to learn the need for continued civil rights activism. And I have sensitivity for where he is as the head of the free world. And it is a thin line to walk. History will decide whether we did it well or not. One thing I am convinced of," he says. Their relationship as black leader and black president "will be a model for how we deal with it in the future, because there is no book yet written about how to do it."

West has attacked me, too, accusing me of selling my soul for "a mess of Obama pottage," and of being one of the biggest "cheerleaders and bootlickers for the president."[48] I have offered principled support for the president in tandem with far more sustained criticism. I have even shared my dissenting views with the president face-to-face at the White House. In *The Center Holds: Obama and His Enemies,* Jonathan Alter recounts a meeting of black figures with Obama in the Roosevelt Room of the White House, where radio host Tom Joyner "began to mix it up with the author Michael Eric Dyson, who wanted the administration to target its efforts more on particular black needs." As I sat directly across the table from the president in an intimate gathering, I disagreed with his universal prescription for helping vulnerable blacks, advocating instead a targeted approach. Alter writes, "Obama jumped in to say he had no problem with Dyson or anyone else disagreeing with him about how to help the needy," but he got upset with "critics who 'question my blackness and commitment to blacks.'"[49] I said nothing to the president that day that I had not said many times over the years.[50]

It is possible to offer substantive, even sharp criticism of Obama without resorting to hateful personal attack, as the work of journalists and public intellectuals Ta-Nehisi Coates and Jelani Cobb proves. Coates has never failed to acknowledge the challenges and complications that Obama faces as the first black president. "I don't wish to minimize the difficulty, rhetorical and otherwise, of being the first black president of a congenitally racist country," Coates wrote in *The Atlantic* in 2015.[51] But Coates has just as consistently taken Obama to task for emphasizing black moral failings while avoiding the structural impediments that plague black life and reinforce black suffering.[52] He argues: "When talking morality in the black community,

Obama has always been very clear. Obama has argued that black kids, specifically, have a mentality which reflects shame in educational achievement. ('I don't know who taught them that reading and writing and conjugating your verbs was something white.') He believes that black men, specifically, tend to be more apt to abandon 'their responsibilities' and act 'like boys instead of men.' He believes that black parents need to learn to 'put away the X-Box' and get kids to bed at a reasonable hour."[53]

Coates maintains that Obama's policy message to blacks lacks the same "targeted specificity" and instead supports a race-neutral progressive agenda: criminal justice reform, infrastructure investment, improved health care coverage, and jobs for low-income neighborhoods. Coates writes that progressives "mix color-conscious moral invective with color-blind public policy. It is not hard to see why that might be the case. Asserting the moral faults of black people tend[s] to gain votes. Asserting the moral faults of their government, not so much."[54] Coates has elsewhere assailed the Obama administration for talking one way to black America—in harsh moral tones—and speaking another way to the rest of the nation by offering public policy to relieve social burdens. Coates concludes that black Americans "deserve more than a sermon. Perhaps they cannot practically receive targeted policy. But surely they have earned something more than targeted scorn."[55]

Jelani Cobb, the author of a fine book on Obama, *The Substance of Hope: Barack Obama and the Paradox of Progress,*[56] has subtly challenged Obama's optimistic measure of racial progress while occasionally decrying Obama's distorted interpretations of black struggle. When the president, in his 2014 speech in Selma, rejected the notion that the Department of Justice report on Ferguson highlighted endemic racism and the belief that little had changed since the sixties, Cobb offered a more expansive historical view.

The President intended to make a useful clarification, yet it's nearly impossible to overlook the fact that the battles in Selma were animated by a local police force empowered to uphold a racially toxic status quo on behalf of a white minority population. Ferguson's is not a singular situation. It is an object lesson in the national policing practices that have created the largest incarcerated population in the Western world, as well as a veil of permanent racial suspicion—practices that many people believe will deliver safety in exchange for injustice. What happened in Selma is happening in Ferguson, and elsewhere, too. The great danger is not that we will discount the progress that has been made but that we have claimed it prematurely.[57]

At times, Cobb has taken note of Obama's calculated historical omissions to make a point, for example, when Obama overlooked the Civil War and armed black self-defense in arguing that black people got their freedom nonviolently.[58] At other times, Cobb has wrestled with the punishing paradox of Obama's historic ascent—and the refuge he takes in the nation's thin veneer of law and order even as it burns in the fires of Ferguson—as when he writes that "perhaps the message here is that American democracy has reached the limits of its elasticity—that the symbolic empowerment of individuals, while the great many remain citizen-outsiders, is the best that we can hope for." Cobb's indictment of Obama's not entirely blameless entanglement in the fiery aftermath of Ferguson stings: "The air last night, thick with smoke and gunfire, suggested something damning of the President."[59]

Cobb also confronted Obama's failure, in his August 2013 speech to commemorate the fiftieth anniversary of the March on Washington, to address the yawning racial chasm that persists five decades after that monumental protest, a failure that riddles the purpose of his office: "That Obama could not—or would not—elucidate his plans to address the intractable realities of race and the economic consequences of those realities . . . calls into question the logic of a black Presidency in itself." Noting that Obama's "tendency to chide black America in public appears all the more cynical when compared with his refusal to point to his own responsibilities to that community as Commander-in-Chief," Cobb directly urged Obama to accountability when he concluded that if "we haven't yet reached that day Dr. King spoke of, then it's required of us to ask the President—even a black one—what he's doing to bring it about."[60]

Just as Coates and Cobb offer examples of balanced critique, Cornel West, in his better moments—especially when dealing with the presidency of Bill Clinton, whom West ardently supported in his first term until political disagreements challenged their relationship—can provide adept presidential criticism. West says that he criticized Bill Clinton for his welfare reform and crime legislation, and thus did not support him for reelection in 1996 as he had in 1992, when he gave speeches for the candidate. Yet when Clinton won again, West says, "I spoke at the inauguration for his second term and was invited back to the White House to discuss a range of issues." West was also part of the group of black thinkers who helped Clinton shape his "mend it, don't end it" speech on affirmative action.[61] West says that despite their differences, he and Clinton "could still talk to each other, still learn from each other, still remain friends."[62] No vicious name-calling, no hurtful epithets in the name of radical purity, just reasonably expressed disagreement that did not preclude friendship and, more important, further support, advising, and conversation, the benefits West objects to when they are offered to Obama.

If black leaders and thinkers have followed different paths in their criticism of Obama, they, and the broader black public, have wrestled with the meaning of Obama's racial identity in a nation where many believed that the first black presidency signaled the end of race. Nothing could be further from the truth.

"INVISIBLE MAN GOT THE WHOLE WORLD WATCHING"

Race, Bi-Race, Post-Race in the Obama Presidency

THE PRIVATE PLANE OF PRESIDENTIAL CANDIDATE BARACK OBAMA taxied down the runway and came to a stop near the SUV that would ferry him, Obama's Harvard Law School professor and then adviser Charles Ogletree, and his campaign aide Ertharin Cousin and me to Obama's New Orleans hotel. Obama was in town to appear at the 2007 Essence Music Festival, where I would endorse him on a pleasant July night before a Superdome crowd of some twenty thousand fans.[1] First we had a frank discussion in the car ride to his hotel about how Hillary Clinton was taking him to school during their Democratic presidential candidate debates. My bluntness sent a jolt of tension through the car, but Obama encouraged me to offer honest criticism and to tell him how he could improve. Obama's penchant for professorial rambling played well in the classroom, I said, but it thudded in the debate forum. Obama ruminated and was sometimes desultory; Clinton was crisp and precise, detailed and deliberate. I suggested he swipe a page from Hillary's debate manual and wed it to a "blacker" rhetorical style. Obama was good-natured about my criticism even though he was fighting a cold amidst the brutal itinerary of an uphill presidential bid.

Later that night I was set to endorse Obama to a crowd that had just listened to rap star Ludacris end a rousing set with his hit "Money Maker." The throng had already sampled the gospel stylings of singers Smokie Norful and Vanessa Bell Armstrong, and had taken in the stylish song-and-dance routine of singer Ciara. Later the audience would hear the immortal tunes of rhythm and blues greats the Isley Brothers and the O'Jays. But now the first black candidate with a chance to make real noise in the presidential primaries would interrupt the musical program for a fifteen-minute speech. Obama bounded onstage and made the effort to appeal to his audience. He took easy measure of the black crowd and referred to the civil rights struggle to draw his listeners closer to him.

"An amazing thing happened in Selma," he said. "People looked at each other and said, 'That's not who we are. That's not what we're about.' And that led to a mighty stream of marchers. Our children are waiting for us to take the same kind of action." Obama made an emotional appeal to the black folk who were still hurting from the horrors of Hurricane Katrina two years before. "After Katrina hit, we had to realize that we were no longer the America we had hoped to be. All the hurricane did was lay bare the fact that we had not dealt with the problems of racism and poverty. The biggest tragedy was that desperate hardship was known here before the hurricane. Poverty double the national average was here before the storm. But here's the good news: America was ashamed and shocked. Our conscience was awakened. We realized that our politics were broken. Suddenly, the curtain was pulled aside to reveal all that."[2] Despite his efforts that night, a journalist noted, some "have doubted Obama's broad-based appeal to the African-American community; at the Superdome, he engaged the crowd but failed to captivate, as have Jesse Jackson and Bill Clinton in the past."[3]

Barack Obama's rapid rise on the national scene raised similar concerns about his appeal to black America and upset a lot of ideas about race. Some critics wondered if Obama were really "black" and what impact he would have on black life and American politics. Obama's biracial identity caused even more confusion. Could a black man reared in Hawaii as the son of an African father and a white American mother shift from immigrant's child to contemporary race man? Many Americans believed that Obama's election meant we would be completely done with race, that we would be living in a post-racial society. If so, how would he relate to the leaders who had forged the path of progress for black America over the last forty years? Obama's biography eventually gave him a boost among skeptical whites even as it sparked the suspicion of many blacks. Some blacks wondered whether this relative newcomer could charm white America, especially the white America he'd said did not exist in the 2004 Democratic National Convention speech that made him famous ("There is not a black America and white America and Latino America and Asian America; there's the United States of America").[4] Most blacks quickly gave him the benefit of the doubt. They knew he'd had to say what he said, maybe even had to believe it, if he was going to help make America whole. Even if there were different Americas, they should not exist, and he was smart enough to say this in a way that did not offend most whites. He was not being dishonest, just a bit naïve, perhaps, or maybe even stubbornly hopeful, but was it not that hope that had often brought a new day in this land?

Obama's high optimism did not keep some black folk from asking if they could trust a man who never got angry or impatient in public. Many more blacks came to believe that his cool demeanor would make him the perfect first black president. Others wondered how any black man or woman who did not sometimes feel anger about the brutal facts of race could be anything but profoundly ignorant of what is really at stake. Or was he just willing to tell white Americans white lies about everything being just fine so they could eat the bread of false brotherhood while many blacks starved from neglect or indifference? Some worried that a half-white man might be halfhearted in speaking for black interests. Before Obama became a supernova in the galaxy of American politics, he was considered an inconspicuous star whose trail brought big questions and grave doubts.

Am I Black Enough for You?

It seems odd that many questioned Obama's literal blackness from the start, as the *New York Times* reported a number of blacks were doing in 2007.[5] It is easy to understand a wrestling with the politics of blackness but not so much with its genetic makeup. Anthropologists have told us for decades that there is no such thing as "race" beyond the meanings we give it in our society. That does not mean that the definitions of race we create and pass along do not dramatically affect our daily lives. It means only that those definitions do not have a strict biological anchor. There is a far deeper interaction than usually advertised between the cultures we make and the chromosomes we inherit. Both play a role in helping us to understand, or to complicate, racial identity. We must keep this in mind when discussing two related but distinct charges made about Obama: that he was not black, and that he was not black *enough*. The former argument is dressed in genes, even if it is also cloaked in social and cultural consequences. The latter charge is a political judgment. Its proponents worried whether Obama would sidestep or misuse traditional meanings of blackness passed on by activists and intellectuals alike. Both of these charges about Obama prove that we cannot just search the soil of science for the meanings of blackness; we must also unearth the ground of our political priorities.

Stanley Crouch and Debra Dickerson, two prominent cultural critics, led the charge of Obama's non-blackness. Crouch and Dickerson have often challenged received wisdom in their politically incorrect views of race, which made it seem likely that they would defend Obama against such charges rather than size him up as one parent and a few genes short of black. Before Obama decided to run, Crouch argued that if he did take the plunge, he would help us understand "the difference between color and ethnic identity." Crouch said that Obama does not "share a heritage with the majority of black Americans, who are descendants of plantation slaves." Thus, Crouch contended, "when black Americans refer to Obama as 'one of us,' I do not know what they are talking about." That is because Obama, despite some brushes with the treatment typically doled out to blacks, "cannot claim those problems as his own—nor has he lived the life of a black American." Crouch maintained that Obama "will have to run as the son of a white woman and an African immigrant," and that should he become our first black president, "he will have to come into the White House through a side door—which might, at this point, be the only one that's open."[6]

A couple of months later, Debra Dickerson, in the parlance of gambling, saw Crouch's racial definition and raised him a qualification or two. She had not earlier had "the heart (or the stomach) to point out the obvious: Obama isn't black." Like Crouch, she argues that "black" describes those folk descended from West African slaves. There is a big difference, she says, between blacks and "voluntary immigrants of African descent." Dickerson thus supports Couch's distinction between color and ethnic identity. To look at them, it is hard to deny that American-born blacks and voluntary immigrants of African descent are both "black" because of color and DNA. But the political and cultural meanings of blackness in this country belong exclusively to American blacks. Dickerson says we "know a great deal about black people" but "next to nothing about immigrants of African descent." Both sides gain in the bargain of our ignorance of the facts. Black Americans get to claim Obama as one of theirs, and whites get to prove they are not racist by embracing a black man—except, in each case, he is not black in the way we usually mean it. Dickerson concludes that Obama "had no part in our racial history, he is free of it. And once he's opened the door to even an awkward embrace of candidates of color for the highest offices, the door will stay open. A side door, but an open door."[7]

Crouch and Dickerson foresaw the "side door" advantage of Obama's non-blackness: black Americans rode the coattails of an African immigrant into political places they would never ordinarily see, while Obama took advantage of opportunities forged by those same blacks who have been denied the benefits he enjoys. Crouch and Dickerson also highlight a paradox: even though Obama may represent black interests, when whites accept him they are not accepting American blacks. What they are endorsing instead is an exotic foreign blackness over an unsexy native blackness. This position represents a reversal of the old saying "Better the devil you know than the devil you don't know." Many whites apparently are willing to take their chances on blacks they do not know rather than the ones they know.[8]

News stories and journal studies detail how white employers prefer hiring West Indian or African immigrants because employers believe they will work harder and more reliably than American blacks. Immigrant blacks do not make as big a stink about poor working conditions.[9] They are also less likely to see race behind unfavorable treatment on the job. Immigrant blacks do not necessarily empathize with the burdens and responsibilities American blacks have borne. But few seem willing to give up the advantages they gain because of black American struggles for equality. Obama has admitted that he has few pieces of racial luggage stuffed with black anger. Often the mere presence of American blacks stirs guilt among whites because blacks symbolize the knowledge of past injustices and evoke resentment over the claim of lingering inequalities.

There also seem to be huge cultural differences and conflicts between black people with lilting accents and black Americans whose tongues give them away as inescapably local. The attraction to exotic blackness often leaves domestic blackness in the dust. It is bad enough to endure such harmful comparisons in private, worse still to suffer such blows while some Africans and West Indians exploit their migrant status in the white world by joining the chorus of attacks on American blacks.[10] But even more punishing may be for blacks to be seduced into voting for figures whose success could be used to reinforce their own degraded value.

It is not difficult to imagine the reluctance of ordinary blacks to embrace a man who had the right color but lacked similar cultural references and social experiences. Every white ethnic community has been unashamed to make that distinction once they got power, as the Irish and Italians did, for example, in Chicago and Boston during the heyday of machine politics. Black voters did not want to get fooled by what Obama looked *like* rather than what he looked *at* when taking stock of the things that matter most to his constituents: the rates of incarceration for the nation's poorest population, the schools black children often stumble through, and the police who too frequently treat black youth with murderous disdain. A leader who might dismiss all of these problems as the fault of black people themselves was either a staunch conservative or the kind of immigrant's son who could not fathom why American blacks would complain about our land of plenty. It would be a sadly ironic gesture of powerlessness for American blacks: raise their voices, cast their votes, and seal their fates. Since they had not had the chance to learn more about Obama, most black folk initially welcomed the friend they knew in Hillary rather than the ally they only hoped they would have in Barack.

There were signs that Obama in fact endorsed an immigrant blackness that was uncomfortable with native blackness, despite his having a foot in both worlds. Obama kept a cool distance from his black colleagues in Congress and rarely bothered with Black Caucus meetings. They seemed to return the favor of the snub by the Senate's only black member. Obama was thrown a fat pitch down the middle of the plate when he was asked whether race played even a small role in the delay of resources to mostly poor black New Orleans residents after Hurricane Katrina struck. Instead of stroking it out the park, Obama was almost inexplicably tentative in his rhetorical swing: he claimed that Bush administration incompetence more than color ruined the day. I say almost inexplicably tentative because, even as a new senator, Obama was calculating how everything he said or did reflected not just on his political present but also on his bright political future, one that few could begin to understand, since he held the cards to the best hand a black political figure in America has ever been dealt. The example of Colin Powell surely played in his mind: you say or do the wrong thing, and the promise that glimmered on you like fresh dew will quickly evaporate and dull your shine.

It was only as Obama proved to black folk that his story was similar to their story that he won them over. On the forty-second anniversary of the Selma voting rights march in 2007, candidate Obama, in his speech at Brown Chapel, faced off against Hillary Clinton, who gave a speech at nearby First Baptist Church, in what the media dubbed a "showdown" for the black vote. Obama directly addressed blacks' fears that he would leave them twisting in the wind because he was not capable of grasping their greatest wishes and deepest pains:

> A lot of people been asking, well, you know, your father was from Africa, your mother, she's a white woman from Kansas. I'm not sure that you have the same experience.
>
> And I tried to explain, you don't understand. You see, my Grandfather was a cook to the British in Kenya. Grew up in a small village and all his life, that's all he was — a cook and a house boy. And that's what they called him, even when he was 60 years old. They called him a house boy. They wouldn't call him by his last name.
>
> Sound familiar?[11]

Obama's speech gave him a critical boost in the polls among black Americans. It would not be the last time he used black vernacular and his personal pilgrimage to woo black voters.

Torn Between Two Lovers

But just as elements of Obama's story earned him credibility, other aspects created doubt about whether he was black enough to truly understand the black plight. Some black people who worried over Obama's blackness focused on how his biracial identity might be used to separate him from ordinary blacks. Not only was there a concern about exotic versus native blackness; there was, too, a fear of how *black-plus* would play against *black-only*.

Obama cites his biracial identity as a bonus in a life whose unlikely success was possible only in America. But Obama has always been slight on the other side of that equation: his triumph here is so unlikely only because America has done more than almost any other nation to exaggerate and exploit racial identity. Race has gained such artificial importance in this country that one group could hog most of the resources for itself and leave all other groups gasping for legal and political air. Obama often takes the knowledge of racial division as his undeclared starting point. By not stating it too much, or too loudly, he can rush past the traumatic memory of race to its positive resolutions. Obama insisted late into his presidency that his biracial identity proved he had the goods to heal the nation's racial wounds. He worked hard to resolve in his body and brain what remains to be fixed in the national body and will—the peaceful coexistence of black and white.

Obama's personal dilemmas have thus been turned to national benefit: they buoy him as he leads a country grappling toward a multicultural future and the slow decline of white dominance. But the question lingers whether Obama's personal dilemmas work against black interests. Obama is willing to underplay evidence of persistent black suffering while promoting a naïvely optimistic view of the depth and pace of racial progress, as he did in the aftermath of the deaths of Michael Brown in Ferguson in August 2014 and Eric Garner a month earlier in Staten Island. Obama's appeal to his biracial identity as a means of resolving racial conflict has at times only served to heighten anxieties and muffle legitimate concerns.

The idea of biracial people makes folk on all sides of the racial divide sweat. The word "biracial" is a vast improvement over other terms that suggest interracial mixture, from "quadroon" to "mulatto," or the slightly sexier "mestizo." But the term "biracial" carries deficits too; when it comes to many ideas that begin with the prefix "bi-"—whether it's biracial, bisexual, or bipolar—America attaches stigma. The most bitter racial abyss in this country savagely separates blacks from whites. Sexual and romantic unions that defy this monumental rift draw attention and outrage. The biracial impulse is doggedly individual; it refuses to give the group ultimate say over the heart or loins. And biracialism rides a paradox into existence: it is often willing to make new kin by ignoring the racial wishes or fears of existing kin. The idea, again, makes all parties nervous.

Obama has written and spoken insightfully about how biracial folk are faced with split racial affinities and competing group loyalties.[12] Because they are members of more than one primary group, biracial people often face an unsettling question: To what tribes and traditions do they rightly or truly belong? It is an inquiry that raises the question of race by having begged off its stubborn restrictions. By giving birth to racial choice from their wombs, the parents of biracial folk cannot help but put flesh on politics. If, as anthropologist Carol Stack argues,[13] black people in America make fictional kin of people not related to them by birth or blood, then biracial people have gone them one better: they create *factional* kin by blending two "races" that are fictions of the social order. Factional kin make both whites and blacks nervous because both "factions" fear a world where lines of purity and authenticity are blurred, though for completely different reasons.

White culture owes a debt to the myth of superiority that most whites in their more honest moments know is absurd. But the fact that the myth is unworkable has not kept it from metastasizing across the culture. The opposition of enlightened whites and their nonwhite allies to the myth has not yet killed it. For unyielding purists, and for lots of rank-and-file folk too, race mixing undercuts the brand and compromises the quality of whiteness.

Many blacks think that peers who hunt for acceptance outside the race really do believe that whites are superior. It is one thing to give in to whiteness at work, or even at play; whites dominate both worlds and have a large part to perform in the employment and recreation of black folk, whether working at Chrysler, cheering at a ballgame, or singing along at a concert. But to surrender to whiteness in love and procreation may suggest that blackness is simply not enough, or worse yet, just not good enough, and that blacks have collectively done what James Baldwin says his father did: believed the horrible things white people said about him.[14] It is clear that the myth of white superiority enjoys an unsavory codependence on its despised twin — the lie of black inferiority.

Some blacks believe that many biracial folk will never be satisfied with the blackness that they will almost always have to learn to make peace with. Even though it seems archaic to say so, the one-drop rule of black blood contaminating white identity still holds sway. Many blacks think that the temptation to despise or avoid blackness, even the blackness contained within one's body, may be too hard for mixed-race folk to resist. In his memoir Dreams from My Father, a compelling and colorful story of his journey to blackness penned before his fame, Obama recalls a conversation he had with a multiracial woman named Joyce who rejected the idea that she was black. "That was the problem with people like Joyce," Obama writes. "They talked about the richness of their multicultural heritage and it sounded real good, until you noticed that they avoided black people."[15] That is why black folk applaud when Barack Obama acknowledges that even though he has white blood, he is a black man. It is also why black folk beam in pride when actress Halle Berry says she is comfortable with her blackness because her white mother told her she would be seen and treated as a black woman. It was also heartening to millions of blacks around the country when New York mayor Bill de

Blasio admitted that he and his black wife, Chirlane McCray, have had many conversations with their biracial son Dante about how to behave when he encounters the police. And it is why blacks wince in painful empathy when Tiger Woods is skittish about his blackness and appears uncomfortable in his own skin as he downplays the fact that others detest or deride that skin.

There is unfair pressure on biracial folk from both whites and blacks who root for one side or the other to prevail. Many biracial people would like to claim both sides of their racial inheritance. Such a desire has its complications. Black people know that blackness is not valued until it proves useful to mainstream culture, while whiteness is always at a premium. Biracial people do not usually get counted as at least half-white, versus half-black, until they do something positive or wonderful to make whites take notice. Before then those same biracial folk were black by default. That may be a lazy way to account for race, but it is a way that has prevailed for more than a century. It is easy for whites to claim Halle Berry, Alicia Keys, or Barack Obama *after* they make good. It is understandable why blacks who love and nurture biracial people might feel betrayed by a sudden and newfound interest in "black" people with mixed blood who, after their fame, become "biracial," meaning half-white, or less-than-black. Some blacks fear that biracial folk who want to drink from all of their racial roots will as soon as possible spit out blackness and guzzle a prized whiteness. White superiority works hard to get a toe in the racial door, even if it is a toe on half a white person "compromised" by black identity.

Obama's conscious effort to be black clearly has a political more than a genetic meaning. Tiger Woods proves that a biracial man who looks black has little say in how the public sees him. But Mariah Carey's success shows how blackness may be suppressed and not seen as a liability in a biracial person who looks white even if she "sounds" black. Black folk often do not acknowledge how biracial kids who want it must struggle for the blackness that black kids take for

granted. Obama, as a biracial child, was born without the compara-
tive black privilege that black kids assume. Black kids have to opt out
of the black world by explicitly denying their roots, like those fair-
skinned blacks who "passed" in earlier generations. Biracial children
must at times actively campaign to prove their authentic black identity. Blacks are born with full racial coverage, while biracial children
are often left uninsured. The hard work of arguing their way inside
blackness may equip some biracial children with more knowledge
of black roots and realities than many black children have. In *Dreams
from My Father,* Obama recalls as an adolescent chiding a black friend
who had not read Malcolm X's autobiography, as the friend retorted,
"I don't need no books to tell me how to be black." Like many studi-
ous black males, Obama writes, "I decided to keep my own counsel
after that, learning to disguise my feverish mood."[16]

As Obama learned his way inside blackness, he did not simply have
to explore the African kinship he shared with his father. He had to
discover as well a black identity his father did not possess, that of an
African American. Obama baptized himself in the black experience
as if he were making up for lost time, which of course he was. He
settled comfortably into the idea that he is a black man in America. It
would be overstating the case to say that Obama did not make much
of his biracial autobiography until he was well known, but it would
not be wrong to say that he has made the most of it since he became
famous. Obama's white mother and grandparents got more airtime
and ink as he struggled to win the white vote and the White House.
New Yorker editor David Remnick has argued that Obama used his
biracial story to try to unify the nation and revive American politics.
"Obama made his biracial ancestry a metaphor for his ambition to
create a broad coalition of support, to rally Americans behind a narrative of moral and political progress," Remnick writes. "He was not
its hero, but he just might be its culmination."[17] Obama seemed to
be saying, "I'm one of you, too," while feeling no need to proclaim
the blackness that he and his audiences, black and white, could take

for granted. It made some blacks, who likewise felt no need to broadcast their blackness, feel right at home; for whites, it did not make
them feel singled out and picked on.

The Politics of Race

Barack Obama has tried to negotiate the clash of race and a broader
domain of identity by insisting that he is "rooted in, but not defined
by" black life. It is a distinction that has become more pronounced
the higher up the political ladder he has climbed. Black life defined
Obama's world when his world was largely limited to his doings and
strivings among black folk. He wanted to venture freely into the bigger world when his opportunities got bigger. Obama calmly insisted
that critics who did not understand his complicated odyssey to blackness would not hold him hostage. He had movingly recorded his
search for his African roots and a conscious and useful black identity
in *Dreams from My Father*. He had engaged deeply entrenched poverty
as a Chicago community organizer and had many blacks among his
constituents as an Illinois state senator. He had joined a church with
an Afrocentric liberation theology and had a broad circle of friends
that included many blacks. As he blossomed in local and national
politics, Obama bravely embodied W. E. B. Du Bois's wish for black
folk to develop themselves in the white world without being spurned
by their own group.

Blacks eventually got over their suspicion of Obama's experiment
in free black identity. They grew to appreciate how he yearned to
crush racial stereotypes and revel in a cosmopolitan American identity. Black folk at their best seek to drink from black roots without
being strangled by them. It made perfect sense to blacks that Obama
would resist being ghettoized as the "black" candidate in his efforts
to win the White House. He stood firm in his views of race. And in
that firmness he taught the nation how being black is not a distortion
of American ideals but rather a celebration of them.

Obama's cosmopolitan views of race served him well as he began to swim in the political mainstream. He made quick work of the "Bradley effect," whereby whites, for fear of sounding racist, say they will support a black candidate but fail to follow through in the voting booth. The most infamous and so namesake example occurred when Los Angeles mayor Tom Bradley lost his 1982 bid for governor of California despite a healthy lead in the polls. When the Bradley effect loomed in a surprising primary loss in 2008 in New Hampshire, where Obama had been pegged to win, he dusted himself off, quickly retooled, picked up steam, and rarely looked back. Of course there were racially tinged worries: moderate white liberals might turn tail at the prospect of a radical shift in the world they had grown accustomed to, a change symbolized by the success of a black candidate whose beliefs they did not yet know. Whites who rode the fence between parties considered stepping onto Obama's turf. But they temporarily had the devil frightened out of them by the vision of Jesus being preached by Obama's longtime pastor Jeremiah Wright, about whom I say more later.

Polls indicated that better-educated whites were more willing than their less-educated peers to give Obama a look the first time around in 2008.[18] It is widely believed that education lessens white bigotry toward blacks and other minorities. The challenge to such a belief can be glimpsed in some well-educated whites who cling to narrow views of race and disturbing beliefs about black folk. But many white Americans proudly embraced Obama as the last best hope of a healthy American politics. Some whites viewed Obama as a heroic progressive, a claim harder to sustain as he got closer to the center so he could get closer to the White House.[19]

A drawback to Obama's adoption of a cosmopolitan black identity was that it led some whites naïvely to believe that his success signaled the end of racism. Former New York mayor and Republican presidential hopeful Rudy Giuliani—who controversially argued in the aftermath of Ferguson in 2014 that black-on-black crime was the reason white cops were heavily posted in black communities—said on air the night of Obama's election in 2008 that his victory would move the nation beyond "the whole idea of race and racial separation and unfairness."[20] Other whites did not quite believe that the death knell of racism had been sounded in a single election. But they did hope that Obama's election would quench racial passion and suspend the need to obsess any longer over race. However the idea got prepared and cooked, millions of Americans feasted on the notion of pushing past race to a post-racial era. That is because blacks and whites are both racially exhausted, but for strikingly different reasons, most of which have to do with clashing stories about the roots and spread of racism, our greatest social malady.

It is hard to believe that the election of a single black man could do away with hundreds of years of bigotry or that his success could wash away the hurt and humiliation that stain the American soul. But post-race rests on an opportunistic paradox: Obama was seen as an exceptional black man who was not quite like most other blacks, but enough like them to relieve white guilt in his election. By voting for Obama, many whites seemed to believe they could collectively clear themselves of the racist charge that had been unfairly hung on them for all the misdeeds of their predecessors. It was as if the sins of the nation were cleansed by his ascension on high, thus bypassing all the ugly business of suffering and death that usually precede such elevation. In fact most discussion of post-race proves we desire resurrection on the cheap: the nation wants quickly to silence talk about race's role in slaughtering black opportunity in the name of selfish group advancement, and then peg its hopes for forgiveness on the body of a figure who absolves it of its transgressions in exchange for his coronation, and his agreement not to talk too much about the original sin of race. Obama has largely kept faith with this scenario.

One of the damaging beliefs held up by Obama's two presidential victories is that there was enough racial offense to go around, and that, in recent times at least, there was equal suffering on both sides of the racial divide. Many whites acknowledge that black folk suffered slavery and Jim Crow apartheid a long time ago. But they also want it known that whites are often wrongly thought of as racist, that their good efforts to end racism are grossly underappreciated, and that whites have been unjustly harmed by affirmative action in the name of black progress. For many whites, the problem is thinking about race at all, not how current thinking about race may be tied to the problems of the past. For these folk there is just as much harm in present efforts to fix what is wrong with race as there is in the existence of old-style racism, which is pretty much the only kind that can be admitted to have existed, since it points the finger away from contemporary whites. Who wants to attend to ugly and entrenched habits that scar the national landscape and grapple with deeply unjust structures that corrupt politics and culture?

A willful immaturity in the Age of Obama stalls contemporary discussions of race and makes it seem that talking about racism is just as bad as racism itself. If the obstacles of the past have been wiped away, the logic goes, then so should the language of race disappear, or be banished. But racism is tricky; it infiltrates language and culture, since those are its original wombs; and it hibernates, and awakens to yawn, and then roar, in institutions like schools and churches, and in structures like the social order and the legal system. Hard-nosed critics who point this out are often blasted for not moving on and accepting the racial success before their very eyes. Obama made this charge during the racial fallout in 2014 from protests against police killings of unarmed black men, saying: "I think an unwillingness to acknowledge that progress has been made cuts off the possibility of further progress. If critics want to suggest that America is inherently and irreducibly racist, then why bother even working on it?"[21] (Perhaps Obama forgot his passionate 1990 introduction of Harvard Law School professor Derrick Bell at a rally in support of Bell's demand that the law school hire a tenured black female professor, where Obama applauded Bell for "speaking the truth" and for his "excellence of . . . scholarship" that led, two years later, to Bell's *Faces at the Bottom of the Well*, a book whose subtitle suggests the nation's persistent plague: *The Permanence of Racism*.)[22] The greatest impediment to progress in this atmosphere is seen as black folk who will not stop talking about race instead of working hard like Barack Obama to succeed. If he can be president, then any black person can do anything. If blacks fail, it is not because there are lingering racial barriers; it is because they are unwilling to take advantage of ample opportunities to thrive.

Black Americans have put forth their own take on such a belief. From the moment Obama was elected president, blacks proclaimed that they had "no excuses" not to dream bigger or to reach higher on account of Obama's success. It was downright inspiring to feel the electricity for excellence circulate even more intensely among blacks because Obama had taken the White House without asking for pity or offering any excuse. This gives the lie to those who claim black people do not get excited over the prospect of individual success. The smartest blacks know they can take heart and draw incentive from Obama's example while admitting that many more Barack Obamas could emerge if they had the opportunity to shine. Barack Obama is not the first black person capable of being president; he is the first black person to get a chance to prove it. When it comes to opportunity it is a two-they street: they, black people, should certainly do all they can to succeed, and they, white people, should do all they can to remove barriers to success.

Comedian Chris Rock brilliantly captures in a 2015 *New York* magazine interview the paradox of progress as a barometer of American race relations in brutally honest terms that Obama could never utter:

> When we talk about race relations in America or racial progress, it's all nonsense. There are no race relations. White people were crazy. Now they're not as crazy. To say that black people have made progress would be to say they deserve what happened to them before . . . There's been black people qualified to be president for hundreds of years . . .
>
> The question is, you know, my kids are smart, educated, beautiful, polite children. There have been smart, educated, beautiful, polite black children for hundreds of years. The advantage that my

children have is that my children are encountering the nicest white people that America has ever produced. Let's hope America keeps on producing nicer white people.[23]

It is easy to see why many blacks have also been seduced by the notion that they have simply got to stop talking about race as much as possible. That is different, by the way, from doing everything they can to make sure that race is not a problem they need to discuss. Many blacks feel that if they are going to succeed, they have got to put the kibosh on race talk and try to better themselves by relying on personal, not social, struggle. This stance rests on common sense as much as conviction, since most whites are fed up with blacks who will not do otherwise. The outlines of black aspiration thus trace the level of white tolerance. Many white Americans say that pure merit determines who gets jobs and school admissions. Such a claim is doubtful when we consider the role of class and legacies in school admissions. And knowing somebody who knows somebody often determines who gets the job—and most of those somebodies are white.[24] Most black folk without these advantages have little choice but to adjust to things as they are and make the best of it. Others take the time to point to obvious flaws in such an arrangement and bear the stigma for having had the courage to say so. And many blacks do both.

It is clear that racial progress has been made in America when courageous blacks and their brave white allies defy the logic of the age to move forward. It is easily forgotten that common sense during the height of segregation counseled against black and white interaction. But the walls of separation fell when folk worked together to prove the unfairness of unearned white privilege and undeserved black suffering. It was terribly unpopular for blacks and whites to denounce the notion of black inferiority and oppose the belief that blacks deserved their lowly spot in the social order. Many blacks were afraid to speak and act on their own behalf; many whites preferred the strong, silent black who endured oppression without complaint.

Most people who identify with the Obama epoch think that, had they been alive at the peak of American apartheid, they would have been on the right side of history. Most blacks think they would have been bravely articulate and would have rebelled against the injustice around them. And most whites feel that they would have certainly cut blacks some slack and tried to do the right thing. It is easy to project our heroism into a past that is not invulnerable to revision. But the past is nevertheless not available to present efforts to change historical outcomes. We find it difficult to face up to the challenges of our present racial era and to determine just how much we should risk to uphold the ideals that got us this far.

Some blacks who have "made it" take refuge in a narrative of racial progress that dismisses the need to discuss race all that much. Obama cannot be spared here; in 2009 he scolded Attorney General Eric Holder for saying that we are a "nation of cowards" because we avoid honest talk about race. Obama said, "I think it's fair to say that if I had been advising my attorney general, we would have used different language," and offered, "I'm not somebody who believes that constantly talking about race somehow solves racial tensions."[25] This made little sense when Obama said it, and it was even less persuasive in the tense wake of two grand juries in Missouri and New York failing to indict police offers for killing unarmed blacks.

Some white liberals think that too much is made of race by black folk who refuse to admit that guilting white folk into line will not work. It is difficult to know the difference between legitimate complaints of too many bullhorns and marches, and white resentment of blacks who look now like pests but who may be seen in the future as prophets. Martin Luther King Jr. had the same kind of mud thrown at him when he walked the earth, yet after he died, he was replanted in the social landscape as a saint and martyred hero. But a big line in the sand has been drawn between those blacks who are willing to get over, and beyond, race and those who are viewed as stuck on race, and therefore stuck in the past. If we are to trust the moral arc of such a narrative, it is the difference between Barack Obama and his Joshua generation of leaders, and Jesse Jackson and his cohort in the Moses generation. If Jackson's generation is bitterly hitched to the race post, then Obama's generation is supposed to be blithely post-race.

Old Heads and New Thoughts

No group of blacks at first proved more cantankerous or troubled by Obama's new blackness and the politics it gave rise to than leaders from the civil rights era. If Obama hailed from the self-proclaimed Joshua generation, the elders from the Moses generation were not yet ready to pass the torch or vacate the scene. And yet it is the Moses generation that has given Obama the fire to embrace blackness and to resist the post-racial myth. This fire is evident in Obama's second book, *The Audacity of Hope,* when he—apple of the post-racial eye—goes sour grapes on the notion of a post-race America:

> When I hear commentators interpreting my [2004 Democratic Convention] speech to mean that we have arrived at a "postracial politics" or that we already live in a colorblind society, I have to offer a word of caution. To say that we are one people is not to suggest that race no longer matters—that the fight for equality has been won, or that the problems that minorities face in this country today are largely self-inflicted. We know the statistics: On almost every single socioeconomic indicator, from infant mortality to life expectancy to employment to home ownership, black and Latino Americans in particular continue to lag far behind their white counterparts. In corporate boardrooms across America, minorities are grossly underrepresented; in the United States Senate, there are only three Latinos and two Asian members (both from Hawaii), and as I write today I am the chamber's sole African American. To suggest that our racial attitudes play no part in

these disparities is to turn a blind eye to both our history and our experience — and to relieve ourselves of the responsibility to make things right.[26]

An older generation of black leaders was shoved into Barack Obama's shadow as the spotlight flooded the new best-selling author and political phenomenon. When shafts of light occasionally fell across their grim faces, only their warts stood out. They had helped black folk survive by their redeeming actions and heroic sacrifices. They had urged black folk to fight back against a society that stole their black humanity and made blacks believe they were not even worth the fight to reclaim it. But now many were distracted by a nagging selfishness; others were plagued by a tendency to define their struggles as "the people's" struggles. Like hip hop artist Kanye West, they let their egos balloon as they did great work. And they could be petty and insanely jealous in scuffling with one another to be Head Negro in Charge.

Many blacks clamored for new blood and new thought. Martin Luther King had been dead for forty years, Jesse Jackson had run a valiant race, and Al Sharpton had pressed his way forward. (Sharpton is interesting because he straddles generations: only seven years Obama's senior, he is a throwback to an earlier era of racial brinkmanship.) But some of the older men would not loosen their grip on resources, and put a vertical hold on emerging leaders. There can be little argument that black leadership has often been a patriarchal party where women and young people are expected to be wallflowers. While many blacks still hungered for the uplifting race advocacy they got from older leaders, whose unapologetic love for the people made blacks willing to tolerate their flaws, it is little wonder that new black leaders wanted to forsake the old ways, and the old men, and give greater account of themselves in running governments and organizations. If Obama was cut from a different cloth, many wondered if he could fit black needs into his pocket as he strode into the mainstream.

Obama was not the only leader to stoke the ire of civil rights stalwarts. Some of them disapproved of Newark mayor, now New Jersey senator, Cory Booker; Washington, D.C., mayor Adrian Fenty, another biracial political wonder; Alabama congressman Artur Davis; and Massachusetts governor Deval Patrick. The schism between old and new leaders is of course universal, but this iteration was made all the more dramatic by Obama's unprecedented ascendance.

Questions about Obama's blackness had obviously migrated from genetics and anthropology to politics: How would he stack up against the best black leaders of the past? Would Obama play ball for blacks when he seemed to be fielding offers from neoliberals and conservative centrists, among other teams? Many black folk gave Obama the thumbs-down at first because they did not want to splurge on another symbolic presidential candidacy. Commentators did not seem to understand that black people were tired of symbols and wanted a real win as much as the next citizen. Blacks had proved their reluctance to believe by their weak support for the 2004 presidential bids of Sharpton and former Illinois senator Carol Moseley Braun. Blacks were not sure at first whether white political forces that did not know black folk, or care about their interests, had propped up Obama.

The black old guard was especially hard to convince because most of them did not know Obama personally. Some who did know him could not be sure that he would have the backs of blacks when it mattered. He had not shown much interest in fellowshipping with many black politicians. Obama sensed from the beginning that he would have more luck with the white mainstream if he was not seen as one of "them," and here one can take one's pick: angry apostle of black grievance, shrill rhetorician of black resentment, or advocate for "black" as opposed to "American" interests. Obama's reading of the white mainstream proved the advantage of his half whiteness: he knew intimately the ways and desires, the fears and frustrations of white folk. His biracial background, as we have seen, made him a kind of racial arbiter who could move shrewdly among blacks and safely among whites, though not without controversy or glitches.

While other black politicians did not mind barking at the big dogs of American politics, Obama wanted to run with them before eventually leading the pack. To take the lead, he would have to zoom to the front of the political mainstream by gaining distance on traditional black politicians. He also needed to tap a natural black constituency that would be his to lose once he could convince it that he was worth the vote. First, he had to prove he had the backing of the white mainstream and could make whites stay with the (Democratic) party that he would lead with their support. Only then would black leaders, who wanted desperately to support a black winner, sign on.

The largest cluster of black political stars at first formed around Senator Hillary Clinton, including civil rights legend and Georgia congressman John Lewis and congresswomen Sheila Jackson Lee from Houston and Stephanie Tubbs Jones from Cleveland. In her 2008 presidential run Clinton had a special rapport with black female politicians, who bonded with the former first lady on concerns of race and gender. Clinton benefited from the considerable goodwill that her husband, Bill, had built among black folk. She also worked quickly to establish her own political footprint in black communities by addressing critical issues from racial profiling to funding for higher education. Hillary rapidly became the front-runner among blacks. She did not lose her lead until Bill Clinton unleashed a series of racially charged inferences during campaign appearances which eroded the Clintons' standing with African Americans. Obama had already pulled off his miracle win in Iowa and was now attracting millions of blacks, forever changing the American political game.

It is telling that Obama could shake the civil rights establishment to its core even though it is hardly a monolithic cast of characters. Andrew Young and Jesse Jackson are both great leaders who are quite different from each other: the former came out of the black middle class to fight the bloody war against inequality, while the latter rose from the ghetto to blunt racism's reach. Yet both men seemed at loggerheads with Obama. Young, on a local Atlanta cable television show in late 2007, humorously handicapped his chances for the nation's top political prize. "I want Barack Obama to be president," Young said, adding, after a dramatic pause, "in 2016." Young argued that Obama lacked the political network that the Clintons could call on to get elected. "There are more black people that Bill and Hillary lean on," Young insisted. "You cannot be president alone . . . To put a brother in there by himself is to set him up for crucifixion. His time will come and the world will be ready for a visionary leadership." Young said that Obama did not have the maturity to mix it up politically and to endure the battles that come with the turf. "It's not a matter of being inexperienced," Young stated. "It's a matter of being young. There's a certain level of maturity . . . you've got to learn to take a certain amount of shit." Finally, in a notoriously bawdy assessment of Obama's fitness for office, Young declared that "Bill is every bit as black as Barack. He's probably gone with more black women than Barack." Young issued a quick disclaimer: "I'm clowning."[27]

Young's comments simmered with angst. Here was an aging freedom fighter apparently overwhelmed by how quickly the world of black politics could shift, with hardly any notice at all, not into a different gear but into a completely different orbit. It was intriguing to watch Young argue that Obama was too young to be president. Young cut his teeth in the movement alongside the most valiant freedom fighter ever, a young man who gained his first fame at twenty-six and violently gave up the ghost before he was forty. Martin Luther King strode toward freedom in a history-changing bus boycott as the pastor of Dexter Avenue Baptist Church in Montgomery. He did not do so as head of First Baptist Church in Chattanooga only because, after tossing his hat in the ring to lead the congregation, he'd come up short, likely because that parish's leaders deemed him too youthful.[28] Young could not seem to make sense of Obama's candidacy and did not seem to realize that he was betraying his own biography in order to make his objections ring true.

There is no doubt that the Clintons had superior black connec- tions at the time, except Young neglected to mention just how skill- fully, and sometimes duplicitously, Bill Clinton had exploited that network, using it to great advantage when he needed to, or discard- ing it at the drop of a hat—or a Justice Department nomination, in Lani Guinier's case. Using Jesse Jackson's annual convention in 1992 to score a hat trick of racial demonology, Clinton crudely ma- nipulated rapper and activist Sister Souljah's words about the '92 Los Angeles rebellion to make it appear that she was a racist urging the death of white Americans; he distanced himself from Jesse Jackson while ensuring the leader could no longer force white Democrats to bend to his will, thus doing to Jackson what Jackson later imagined doing to Obama—cutting off his power at its source; and he sig- naled to white suburbia that it mattered just as much as urban black America. His actions were at once crafty and cynical; they perfectly forecast Clinton's stormy efforts to rain down his subversive magic on Obama on behalf of his wife in the 2008 campaign. (Times have changed: as gifted a racial alchemist as Clinton was, he became, dur- ing Obama's first presidential run, wan and disposable among the masses of blacks he had once effortlessly charmed.)

Young was plain wrong that Obama had not built a network, or at least was not quickly doing so, not just among black folk but among as broad a sweep of Americans, from crabgrass communities to cy- berspace outposts, as any black, or for that matter any American, politician had ever canvassed. Obama roped in millions of young black folk even as older blacks increasingly heard his siren call. If Young could be forgiven for not knowing that these people would support Obama, he and other black politicians probably could not forgive themselves for having failed to inspire such a voter-rich dem- ographic to come out to the polls for the first time.

Young's contention that Bill was as black as Barack really says more than it seems to on the surface. To say that a white man is as black as a black man is to say that the black man is not as black as the white man, since the black man should have been blacker than the white man because he got a head start at birth. In the addled game of racial authenticity, there is no such thing as a black man winning by merely tying a white man. The tie goes to the runner, and while Obama was just stepping up to the plate and taking a swing at being black in public, Clinton had been on base for a while with blacks. According to Young, he had been scoring as well with black women, which proved his blackness, though such a sexist measure of authenticity might seem more likely to spill from the lips of rapper Young Jeezy than from Andrew Young. It may have been a sly and sexually denigrating way of calling a spade less than a spade; Obama did not get the benefit of racial doubt because of his political pedigree. And whether Young intended it or not, Obama's biracial roots, or at the very least his complicated racial journey, was in question.

Young's painfully adolescent words could not quite be believed as mere comic bluster. As the old black folk say, one can be "crackin' but factin'"; in other words, jokes do not completely disguise the truth. At times Obama did not take such not-black-enough accusations lying down, as when he rebuked those who challenged his role in black politics during his first Selma speech in 2007 at Brown Chapel: "So don't tell me I don't have a claim on Selma, Alabama. Don't tell me I'm not coming home to Selma, Alabama." The simple truth is, there was a new sheriff in town, and neither Young nor his fading gerontocracy had grandfathered him in. Young and company lost their bearings; old racial landmarks receded, and up sprang cyber-links that heralded a new black political energy and a true rainbow coalition.

One might think that the rainbow sign would endear Obama to fellow Chicagoan Jesse Jackson, but from the start there seemed to be an uneasy alliance between Jackson's Obi-Wan Kenobi and Obama's Luke Skywalker. The Force was immortally summoned by Jackson's historic presidential campaigns in '84 and '88. Jackson made the very idea of a *black* president reasonable in many quarters where the belief had barely existed at all. Jackson tirelessly evangelized the nation with his circuit-riding defense of poor blacks and whites, correcting the myopia that segregated politics by pigment. Jackson built bridges into Latino communities and energized a flagging labor movement to fuel the most viable progressive movement in a generation. His unforeseen success in 1984, along with the Democratic Party's drubbing at the polls that year, and Jackson's insistence on greater transparency in the nominating process, led to rule changes that linked delegate count and electoral strength. Convention delegates would now be awarded according to a candidate's share of the popular vote in state primaries. While Jackson got only 20 percent of the popular vote in '84, the change of rules he inspired doubled his percentage of delegates in '88. That change, along with the creation of majority-minority legislative districts that owed a debt to Jackson's relentless voter registration drives, opened the way for Obama's victory twenty years later.[29]

Jackson endorsed Obama early enough, but he had little choice to do otherwise. If he had not signed on, he would have been viewed as a "hater" who could not stomach being eclipsed by the first black candidate with a superb chance of taking up residence at 1600 Pennsylvania Avenue. Jackson had been cordial to Obama as he gained momentum on the local scene. Obama at first was appropriately deferential to Jackson as black America's Big Kahuna. Jackson felt the tug of his own blood to pitch camp with Obama. Jackson's daughter Santita was a close friend of Obama's wife, Michelle, and godmother to the Obamas' older daughter, Malia, and Jackson's congressman son Jesse Jr. and Obama shared a brand of politics more subtle than Jackson's brash tactics of confrontation. Jackson *fils* had been appreciative of his father's pioneering path, but he barely attempted to disguise his frustration with the lack of accountability he spotted in old-style black leadership. This view brought the younger Jackson and Obama into rough sync; Jesse Jr. served as a national co-chair of Obama's presidential campaign. Congressman Jackson admonished and cajoled his colleagues to Obama's side in getting out the black vote.

No one was fiercer in defending Obama against the black old

guard than Jesse Jr., although it grew increasingly uncomfortable to see him sparring publicly with his father. When Jackson *père* chastised all the presidential candidates other than John Edwards in a newspaper column for failing to address the plight of black America, including the Jena Six—a group of black teens charged with attempted murder in Jena, Louisiana, for beating up a white student after three nooses were found hanging from a tree on school grounds in 2006—Jesse Jr. responded with a letter published in the *Chicago Sun-Times* titled "You're Wrong on Obama, Dad." Claiming to have witnessed "Obama's powerful, consistent and effective advocacy for African Americans," Jesse Jr. argued that Obama "is deeply rooted in the black community, having fought for social justice and economic inclusion throughout his life." He said that on the campaign trail, as in the U.S. Senate, and in the Illinois state legislature before that, "Obama has addressed many of the issues facing African Americans out of personal conviction, rather than political calculation."[30] The last phrase of the last sentence no doubt stung a bit more because it took aim at his father's perceived opportunism. It did not help that a South Carolina newspaper reporter alleged that Jackson *père* had accused Obama of "acting like he's white" when Obama offered a tepid written response to the charges against the Jena Six by saying simply that they were "inappropriate."[31]

The ugliest blow of all came when Jesse Jr. responded angrily after the elder Jackson, ticked because he believed Obama had offered moral lectures to black folk while extending hope and promises to other communities, uncorked his disgust at Obama's appearance on Father's Day in 2008 at a black church in Chicago where the presidential candidate took black fathers to task in a widely cited campaign speech. Before he went on air at Fox News Channel to discuss health care with fellow panelist Reed V. Tuckson, a health executive, a rolling camera and live microphone recorded the elder Jackson's whispered desire to "cut [Obama's] nuts off" for "talking down to black people" while also capturing the slicing gesture that accompanied his wounding words. Fox host Bill O'Reilly broke the story a few days later and then covered it on his evening broadcast. Jesse Jr. was irate, knowing that whatever negative thing his namesake did reflected poorly on him.

"I'm deeply outraged and disappointed in Reverend Jackson's reckless statements about Senator Barack Obama," Jesse Jr. said in a press statement. "Reverend Jackson is my dad and I'll always love him. He should know how hard that I've worked for the last year and a half as a national co-chair of Barack Obama's presidential campaign. So, I thoroughly reject and repudiate his ugly rhetoric. He should keep hope alive and any personal attacks and insults to himself."[32] It was an instance of a proud but aging king of the jungle having to hear the roar of a new pride of lions. Sadly, Jesse Jr.'s ferocious defense of Obama did not ultimately reap much dividend for the younger Jackson. It could even be said that Obama hung Jesse Jr. out to dry at a critical juncture in Jackson's career. Not only was Jesse Jr.'s name not on a list of suitable candidates to fill out president-elect Obama's Senate term, but also Obama failed to offer a single note of public consolation to Jesse Jr., or to vouch for him in any way, when he faced severe legal questioning about his role in seeking to replace Obama in the Senate. Obama remained silent during the younger Jackson's legal ordeal, which eventually led to his 2013 conviction for wire and mail fraud and a thirty-month federal prison sentence.[33]

Obama realized that the fallout over the senior Jackson's slashing statement was a big win for him. Some younger black politicians and activists viewed Jackson as still valuable, if also hopelessly old-school. Many whites viewed him as part of the problem rather than the solution to the race issue, although a number admired him. The Obama camp could not afford to write him off; criticizing Jackson in public would be disrespectful and costly among those who highly regarded Jackson as well as Obama. Jackson's intemperate remarks gave Obama and his campaign the cover they needed. Clearly they could not put him to work on the campaign trail now lest he go off and say something that did not capture Obama's nouveau black politics or represent his foreign policy views.

Jackson's and Obama's clashing responses to the not-guilty verdict in the trial of white New York City policemen accused of murdering a black man, Sean Bell, the day before his November 2006 wedding in a hail of fifty bullets in New York is instructive about the former concern. Jackson called the shooting "a massacre" and said the verdicts were "a travesty of justice," while Obama said, "The judge has made his ruling, and we're a nation of laws, so we respect the verdict that came down."[34] As for the latter concern, when Jackson allegedly said in an interview that Obama would reverse decades of American foreign policy that put Israel's interests first in the Middle East, the Obama camp shot back that Jackson was not a spokesman for the campaign and reassured Israel that its status would be supremely safe with Obama. Jackson was officially on the roster as an early endorser, but he was relegated to the injured reserve list, never seeing any playing time, a snub that chafed the civil rights superstar.

Without lifting a finger, the Obama camp gained another advantage from the reverend's cutting comments: Jackson's public embarrassment almost guaranteed that he would tone down even his aboveboard criticisms of the candidate. If he had had the chance to criticize Obama openly without risking the wrath of black America, Jackson would not have had to whisper his disapproval. The presidential candidate got a boon in Jackson's offer of a severed package: Obama need not publicly humiliate the wounded giant; he simply had to let him cook in his own juices. His gracious acceptance of Jackson's apology for his "crude and hurtful" remarks was back-ended by Obama's defiant insistence that he would continue to demand responsibility of black fathers. Jackson's powerful criticism of Obama's moral rampaging in black communities was smothered in the blizzard of bad press over Jackson's ill-chosen words. Obama thus escaped responsibility for his one-sided demand for responsibility. It would hardly be the last time that Obama issued an imbalanced call for racial responsibility.

If Obama only reluctantly and cagily tipped his hat at first to civil rights heavyweights like Jackson and Sharpton, he carefully choreographed a shrewd two-step: he leapfrogged past them to speak often of Martin Luther King Jr., channeling the dead hero's oratorical optimism and embracing the inspiration behind the old guard leaders while avoiding association with their perceived stains or stunts. It was only after Sharpton offered surprising support to Obama's campaign, largely by not becoming a liability and remaining relatively silent on the sidelines, that he later became the president's favorite black leader.

Obama would not have made it to the mountaintop, however, if a great cloud of witnesses had not cheered him on, and if noble leaders and flawed emissaries of racial redemption and political change had not paved the way. In short, Obama weaved in and out of the same tradition that produced Jackson and Shirley Chisholm, and before them, figures like Booker T. Washington, W. E. B. Du Bois, Marcus Garvey, Mary McLeod Bethune, Elijah Muhammad, Dorothy Height, Malcolm X, and Dr. King. If Obama's rise did not signal the end of these leaders' importance, neither did his ascendance mean the problems they confronted had gone away. It quickly became clear that Obama's election did not end racism, the civil rights movement, the struggle for black equality—or the careers of Jackson and Sharpton and thousands of other local and national leaders. Obama's success was fueled in large measure by the freedom urges of black folk that were expressed, but not exhausted, in Obama's victory. The post-racial ideal hardly encompassed black aspirations or Obama's historic achievement.

Post-Racial, Post-Racist, and the Fictions of Race

A post-racial outlook seeks to ignore, or destroy, race; a post-racist outlook seeks to destroy racism. There is more than semantics in the balance. Race may be a fiction, but like all good fiction, say, *Moby-Dick* or *The Color Purple*, it may be true even when it is not real. To be sure, what is true or false about race is often determined by what is at moral and social stake for us; those interests shape what stories we tell ourselves about the condition and worth of our society. Race is intimately yoked to our values and self-image as a culture. The idea of race can be tracked up the mountain of philosophy, and traced into the depths of politics and power where fear and fantasy mingle.

Race has surely been used to beat black folk into submission; through its power, they have been seduced into cooperating with their self-destruction. But race has provided positive terms for black folk to forge solidarity and to change their lives through concerted action. Nowhere has this hope been more powerfully expressed than with the election of Barack Obama as president. Blacks have also taken hold of race to tell the truth about their existence beyond the force of oppression. There are untruths that hamper our understanding of how race operates. What is not true about race is the alleged biological input that separates one group from another; what is true about race is that culture and identity are invented in space, and build, or erode, over time. What is not true about race is that the intelligence of members of a group is innate and tied exclusively to a group's genetic structure; what is true about race is that different qualities in a group are born when opportunity marries environment. What is not true about race is that the information we gather about groups under the imperatives of stereotype is reliable; what is true about race is that stereotypes are a hazardous way to find out about cultures, and are poor substitutes for direct experience and wise reflection.

Some have questioned whether race is a useful cultural fiction, or if, like other unstylish beliefs, it should be discarded. Fiction may not be the problem; the trouble may be our storytellers and the stories about race they are narrating. Race was undoubtedly invented to control groups and to justify others' stake in their intellectual and moral inferiority. But the subjects of race have expanded and spoken back. It may be helpful to understand race not only as fiction but also as the language that fiction speaks. The problem is not the English language but what has been made of it—to what ends it has been put and for what purposes it is spoken. Both the NAACP and the Ku

Klux Klan use English to express themselves. What is wrong or right is the moral vision being expressed, not the language through which it is expressed.

The stories and storytellers of race can certainly stand some change. Barack Obama's election promised at least to open up space for new narratives of race to be shared by new narrators. Those who have had the greatest amount of influence in our culture have anchored the story of race in dominant whiteness: white perspectives are taken for granted, white interests are given priority, white identities are seen as normal, white intelligence is seen as superior, and white morality is seen as universal. The white world was good and natural, and when things got out of line, whether because of moral decay or social pathologies, things were either corrected or purged. All other identities that took root in the white world—either because they had been conquered as the white world violently expanded or because they were forcibly transplanted—were viewed as fatally deficient and substandard. The only fix was to become white, or at least as white as possible, in behavior, outlook, and desire. To wish otherwise, that is, to wish to define oneself outside the ambitions of whiteness, was to risk the wrath of white folk who were willing to "save" nonwhites from their God-forsaken plight by killing them into shape—or, to put it crudely, to achieve redemption through genocide.

For black folk and others, dominant whiteness reflects the distorted, defective uses of race, which is the very definition of racism. The problem is not race per se, especially since race cannot be easily, or cleanly, separated from the cultures black folk and others carved from their given circumstances in the white world. Such racial self-understanding and self-definition form the baseline of black cultures. That surely cannot and should not be destroyed, or got beyond, or over, or be forced to disappear or evolve into an amorphous, race-less entity. The problem, therefore, is not race; the problem is *racism,* or the artificial narrowing of racial identity to the size and shape of dominant whiteness. Black figures from Frederick Douglass to Barack Obama have challenged the narrative of dominant whiteness, which reads like so many bad novels of racist ideas whose narrators have corrupted the language of race to express their ingeniously destructive plots.

Ironically, the success of black folk and others in the fight against dominant whiteness has caused race, and eventually racism, to be blamed on them. The very people who turned a liability into an advantage by redefining racial terms, and improving their destinies, have taken the rap for race. One of the privileges of whiteness is the ability not to appear white at all, but to be seen simply as "human." It is as if to say to black folk—who claim that the point of all their protests, from Selma in 1965 to Staten Island in 2015, is to be treated like human beings—"Hey, we've been here all along waiting on you." Whiteness is also a way of turning back black appeals to race—although the appeal to race was forced on them by white society—to suggest that blacks would all be better off by giving up race talk and racial history, and acting as if all persons are now equal. It is the brilliant courtship of plausible deniability and historical amnesia. It is black folk who are made to look obsessed with race. It is black folk out to defend themselves against dominant whiteness who are made to appear racist or, in the ultimate linguistic contortion, reverse racists. Thus when most whites hear "race," they see black. Post-race is really black disappearance.

As it turns out, many advocates of a post-racial world do not want to get beyond race; they want to get beyond blackness. Post-racial sounds forward-looking, but in truth it harks back to the troubling wish for Negro removal that fueled the movement to send blacks back to Africa in the nineteenth century. The same forces arguably lay behind so-called urban renewal in the twentieth century.[35] The fantasy that blackness can somehow be done with, overcome, gotten rid of, quenched, quarantined, cordoned off, or finally resolved is what really lies behind the post-racial ideal. That may explain the desire of many Americans for Obama to be non-black, or at least only half-black.[36] Some believed that Barack Obama had to stop being a black man to govern effectively; but America must continually overcome its brutal racist past to permit his gifts, and those of other blacks, to shine. Opening the door to Obama's presidency was a big step in the right direction because black folk should not have to stop being black to be seen as fully human and completely American.

Perhaps if we compare race to gender we might see the point. Enlightened women and their male allies do not want this to be a post-female society. We want this to be a post-misogynist society, a post-sexist society, and certainly a post-patriarchal society. We do not want women to stop being women. We want men and women to get past ill-informed beliefs about women. We want to overcome sexist behaviors that slight women's humanity and trump their social equality. The same outlook should apply to race in our first black presidency.

Barack Obama's presidency, however, has hardly put a dent in the forces that pulverize black life: high infant mortality rates, high unemployment, atrocious educational inequality, racial profiling, and deadly police brutality. That does not mean that Obama's presidency bears little symbolic value; there is huge meaning for his people, and the nation at large, in the fact that the leader of the free world is a black man. It shows we have matured as a nation. It proves that we can look beyond color to see character and credentials. Obama's victory may one day spur us to slay the dragons of racism and inequality that continue to stalk the national landscape. But it does not mean that we have arrived in the racial Promised Land or that we are done with blackness. There is a new blackness in the nation that must be praised for its advances and scrutinized for its contradictions and weaknesses.

★⫽⫽⫽ 3 ⫽⫽⫽★

BLACK PRESIDENCY,
BLACK RHETORIC

Pharaoh and Moses Speak

PRESIDENT OBAMA REGALED THE AUDIENCE IN THE SPRING OF 2012
with his humor in what has to be one of the most enjoyable roles
for any commander in chief: stand-up comedian at the annual dinner
for the White House Correspondents' Association. I had watched
the event in previous years on my television screen, but the live version was far more entertaining. Professional comic Jimmy Kimmel
nervously rushed through his jokes and even stepped on some of
his own lines. Obama, by contrast, was smooth and poised, confident that his zingers would find their mark. His swag quotient was
high that night as well; he insisted that it was no fluke that he'd offered a pitch-perfect rendition of a line from Al Green's R&B classic
"Let's Stay Together" at an Apollo Theater fund-raiser three months
earlier.

Obama's version of the ordained minister and soul legend's tune
had gone viral in black communities. It demonstrated the president's

effortless embrace of black culture despite criticism that he had been keeping blackness at bay. After Obama drew house-rocking applause from his largely black audience at the Apollo, he addressed Green, who, along with songstress India Arie, had sung at the affair. "Don't worry, Rev," Obama said. "I cannot sing like you, but I just wanted to show my appreciation."[1]

At the correspondents' dinner Obama again flexed his musical muscles, this time by displaying an aficionado's grasp of hip hop culture.

The setup for Obama's hip hop coolness was a perfect storm of dissing conspiracy theories and embracing black cultural signifying. "Now, if I do win a second term as president," Obama said, teasing his audience, "let me just say something to all my conspiracy-oriented friends on the right who think I'm planning to unleash some secret agenda." He paused for a few seconds, then hit them with an affirmation of the conservatives' worst nightmare: "You're absolutely right!" Obama had a mischievous look on his face as he lowered his voice in faux-ominous fashion to clinch the conspiratorial conceit. "So allow me to close with a quick preview of the secret agenda you can expect in a second Obama administration. In my first term I sang Al Green," the president deadpanned. "In my second term, I'm goin' with Young Jeezy!"[2] He accented the second syllable of Jeezy and stretched it out a bit in dialectical deference to black street pronunciation, so that it sounded like Gee-*zeeee*. The audience roared its approval of his self-confident reference to rap royalty, as much out of the desire to be hip right along with him as to reward his sure-handed grasp of hip hop culture. Everything about Obama's linguistic charisma was on display that night: humor, signifying, self-deprecation, pop cultural references, vernacular verve, and effortless code-switching.

To understand what Barack Obama does with language—how syllables sound and words crackle in his mouth, and how his voice elevates or dips, and the sentences simmer, or sometimes sing—one cannot simply turn to theories of rhetoric or listen to other gifted presidential communicators like John F. Kennedy and Bill Clinton.[3] As much praise as Obama has justly received for speaking in a way that does not assault the white eardrum or worldview, his rhetoric is firmly rooted in black speech, whose best rhetoricians marry style and substance to spawn a uniquely earthy eloquence.[4] Obama's oratory hums with the kind of speech he heard for years from black preachers in the church pulpit. His former pastor Jeremiah Wright took it on the chin with the charge that he was an anti-American racist when his words were taken out of context, but his sermons fed the sacred appetite of our first black president. If one likes Barack Obama, one has got to like some parts of Jeremiah Wright. As I lay out later in this chapter, the culture's ignorance of the black church helped fan the flames of controversy that lapped at Wright's words.

Obama has relentlessly channeled the legendary life and outsize oratory of Martin Luther King Jr. Most observers believe that Wright and King have little theology in common and that their rhetoric and moral aims are in deep conflict. The truth is that Wright and King share the same prophetic outlook and read politics and history through a similar biblical lens. It is also true that Obama and Wright emphasize King's split mind on race. King's shifting views can be sorted by chronology and color: King at first thought whites could be persuaded to change, but he grew to believe in the later years of his life that transformation had to be forced. In contrast to his appearances before white audiences, the leader was far more angry and transparent among blacks. The earlier King is a hero to Obama's heart and tongue. As a presidential candidate Obama benefited when his image appeared beside the martyr's on posters plastered throughout black America.

A central contradiction loomed from the start: Martin Luther King Jr., the leader of his people, prophesied social change as an American Moses; Obama sought to become the nation's leader, America's Pharaoh. The conflicting ends of each man have been amplified in the way they used black speech to advance their social and political visions. The tension between the two styles of leadership came to a head in Obama's forced confrontation with his former pastor. He at first conditionally affirmed Wright by carefully endorsing a limited prophetic ambition while criticizing Wright's particular prophetic style and reach. It was a hazardous balancing act between competing moral and social ends that exposed a monumental ethical shift in black life: the political, at least momentarily, trumped the prophetic. Obama eventually parted ways with Wright, but not before delivering one of the most eloquent, and one of the most troubling, speeches on race ever delivered by an American politician.

The Black Religious Wellsprings of Presidential Oratory

The way Obama speaks owes a debt to the cadence and the colloquialisms, and the humor too, of the black pulpit. Consider a campaign speech he gave in 2008, in Sumter, South Carolina.[5] Obama was a bit peeved at how his ideas about taxes were being misrepresented by his opponents in the Democratic presidential primaries, and at how his acknowledgment of Ronald Reagan's ability to get Democrats to vote Republican had been unfairly twisted into praise for the Great Communicator's ideas.[6] Addressing a largely African American audience, Obama let loose with the black tradition known as signifying—in which the speaker hints at ideas or meanings that are veiled to outsiders.

"They're trying to bamboozle you," he said. "It's the same old okeydoke. Y'all know about okeydoke, right?" he asked, as the audience erupted in laughter at his comic timing and vernacular charm. Keeping up the humor, he scorned the idea that he was a Muslim, insisting, in a spurt of black English, "I've been a member of the same church for almost twenty years, prayin' to Jesus—wit' my . . . Bible." And he repeated his theme of political trickery: "They try to bamboozle you. Hoodwink ya. Try to hoodwink ya. All right, I'm having too much fun here."

Ironically, in style and substance, Obama's flight of rhetoric recalled, of all people, Malcolm X, at least the one portrayed in Spike

Lee's biopic, who says in a memorable speech from the film: "You've been had. You've been took. You've been hoodwinked. Bamboozled. Led astray. Run amok."[7] Obama was making a risky move that played to inside-group understanding even as he campaigned in the white mainstream. While denying that he was Muslim, he fastened on to the rhetoric of the most revered Black Muslim, mimicking his tone and rhythm beat for beat. Obama's seamless code-switching was a gutsy, playfully defiant gesture, rife with black humor and tropes. But if you were not familiar with black culture, most of what he said and how he said it went right over your head—and beyond your ears.

For another example, take the 2012 White House Correspondents' Association Dinner speech, in which Obama, after teasing the crowd about singing the music of Young Jeezy in his second term, noted that his wife approved of his choice. Turning to First Lady Michelle Obama as she smiled broadly and signaled her affirmation at the head table, Obama offered a humorous ad lib: "Michelle said, 'Yeah!'" After the laughter rippled across the room, Obama continued, "I sing that to her sometimes." Michelle Obama bent her head and blushed at the public confession of private affection. President Obama flashed his famous pearls for the crowd in the hotel and across the globe.

Obama's gesture was replete with meaning. It was more than an inside joke, a fetching moment of affection between husband and wife played out for the world to see. We cannot forget that not all such inside exchanges had been fondly received by the outside world; during the '08 campaign, the couple's infamous fist bump, a love tap of camaraderie and an affectionate gesture signifying "We're in this thing together, babe," made the cover of *The New Yorker* and earned the enmity of even the limousine liberal set as a sign of some kind of kinky black—and for some, terrorist—code.[8] Now Obama may have been imparting an even more veiled message to hip hop's constituency in the hood that America's first black president, despite the claims otherwise, had not forgotten about them or their needs. The first thing Obama suggested about his administration's second term, joking or not, was an explicit embrace of hip hop by the commander in chief. The humor could not ultimately diminish the spotlight Obama gave to the culture.

Jeezy was not simply a protégé of Obama favorite Jay Z; he was the rapper who famously touted black pride during unofficial inauguration ceremonies in 2008 with his anthem "My President Is Black," a tune he originally recorded with rapper Nas six months before Obama's election. Given the racial lay of the land, Obama could hardly embrace all of Jeezy's output without courting blowback and complication. It should be remembered that after the national tragedy of Trayvon Martin's death in 2012, Obama could not even say that if he had a son, he would look like Martin, without the right wing venting its caustic rhetoric.[9] In a harmless context where even plausible deniability seemed unnecessary, Obama returned the favor to Jeezy, saying, in essence, "Yes, beyond narrow views of race and blackness, and beyond the hate of the ignorant, your president *is* black."

To get a clear glimpse of his considerable rhetorical skills, hark back to the first time Obama caught our attention—with the keynote speech at the 2004 Democratic National Convention.[10] Embracing Illinois senator Dick Durbin, who had introduced him, with a black man's right-hand-dap-then-shake-and-left-arm-hugging-the-back swooping gesture, the little-known Illinois senatorial candidate waited for the strains of Curtis Mayfield's "Keep on Pushing" to subside before he rode the rhythm of hope straight into the hearts of America. On that occasion he obeyed the black preacher's dictum: "Start low, go slow, rise high, strike fire, and sit down."

After telling the story of his biracial roots, applauding the aspirations of ordinary Americans, and praising the virtues of democracy, all in measured tones, he built steadily and rhythmically, with shifting cadences and varied registers, to a climax that exploded in lines of warning to cynics who would divide the country into blue and red states, thinking that they had color-coded the country's ideological divorce: "Well, I say to them tonight, there is not a liberal America and a conservative America; there is the *United States* of America.

There is not a *black America* and a *white America* and *Latino America* and *Asian America;* there's the United States"—a slight pause, on beat, for more emphasis still—"of America." (I have questioned the motives behind Obama's "not a black America and a white America" assertion, but there's no questioning his delivery.) Obama capped his oration with a device often used by black preachers in backwoods and urban pulpits alike: anaphora, or repeating the same word or phrase at the beginning of successive sentences. "I believe that we can give our middle class relief . . . I believe we can provide jobs for the jobless . . . I believe that we have a righteous wind at our backs."

Martin Luther King Jr. relied heavily on anaphora, especially in the "I have a dream" refrain of his most famous speech. Obama used it to brilliant effect after he won a surprise victory in the Iowa caucuses against Hillary Rodham Clinton: "They said this day would never come. They *said*"—this time he stretched the word in sweet glissando, as one effortlessly glides from one pitch to another, which in Obama's case was higher, but in other famous cases is lower, as with boxing announcer Michael Buffer saluting a heavyweight champion of the world, scaling invariably down on the word "world" as if he were sliding on the ropes from a knockout punch—"our sights were set too high. They said this country was too divided."[11] But of course it was not, because a black man had pulled off an unlikely upset in a nearly all-white state.

Like King, Obama can be heard reversing the strategy of anaphora and instead milking the pleasures of epistrophe, in which a speaker repeats the same word or phrase at the end of successive sentences. A mere five days after his win in the Iowa caucus, after barely losing the New Hampshire primary to a surging Clinton, Obama rallied his troops and thrilled the nation. It all came down to a bit of verbal alchemy conjured by his young speechwriter Jon Favreau, in a phrase pinched from United Farm Workers leaders Cesar Chavez and Dolores Huerta that Obama trumpeted with stirring, even defiant confidence at the end of several serial clauses: "Yes we can." Obama argued that this creed "was written into the founding documents that declared the destiny of a nation: Yes, we can." On he went for several more sentences, and then ended with "three words that will ring from coast to coast, from sea to shining sea: Yes, we can."[12]

This is not to say that Obama has mastered every type of public communication. He beat John McCain in debates leading up to the 2008 general election, but Hillary Clinton often bested Obama and his male colleagues in their Democratic primary debates. Moreover, Obama's speech is sprinkled with hiccups and hitches, "ahs" and "ums"—a verbal tic encouraged, no doubt, in academe, where one learns to be extremely cautious, reluctant to offer sweeping statements without justification, and where arguments sometimes die the death of a thousand qualifications.

But there may be more to Obama's "ahs" than meets the ear. Obama's speech, like that of other blacks, may be pressured by his awareness that what he says will be nearly infinitely parsed. To be sure, that is true for most politicians. But it is even truer for a black politician, even one who, like Obama, has gained fame for his ability to talk. We all found this out when, at the end of a press conference on health care reform early in his first term, President Obama let on that he thought the Cambridge police had "acted stupidly" in arresting Harvard professor Henry Louis Gates in his home. Those two words ignited a week's worth of national controversy and conversation on race.

We got a glimpse of the surprise at black verbal facility in 2007, when presidential candidate Senator Joe Biden made his gaffe about Obama being "the first mainstream African American who is articulate and bright and clean and a nice-looking guy."[13] It was not the fact that Biden called out Obama's ability to talk that raised eyebrows among blacks; it was the fact that in his view, and perhaps that of millions more, Obama was the *first* "articulate" and bright African American to chase the presidency. Behind his praise of Obama was an assumption concerning the vices of black speech, a suspicion of its ability to be eloquent or analytical.

Obama took no offense at Biden's words—after all, he later chose

Biden to be his running mate—but called them "historically inaccurate." Black presidential candidates including Jesse Jackson, Shirley Chisholm, Carol Moseley Braun, and Al Sharpton, he said, had given "a voice to many important issues through their campaigns, and no one would call them inarticulate."[14] Obama understood that his gifts made him an extension of, not an exception to, the group. I explore later in this book how much Obama's insight about speech translates to the world of politics.

Obama's speechmaking and oral signifying reflect the beauty and power of black rhetoric—its peculiar rules and regulations, its sites and sounds, its labyrinth of complicated meanings, its bedazzling linguistic variety, its undulating cadences, and its irreverent challenges to "proper" grammar. As linguists Geneva Smitherman and H. Samy Alim cogently argue, Obama would not be president of the United States of America without being a past master of black (American) rhetoric.[15]

To say that he spoke his way into office is not to reduce Obama's achievement to his ability to speak "standard" English. It is a lot more complicated than that. Language is as big as politics, as large as the geography that encompasses the American populace and the demographics that dot the national landscape. Obama's achievement, likewise, is bigger than adding up the parts of speech he uses. It is also about understanding the cultural traditions that feed and shape his linguistic appetites. It is about knowing the racial practices against which that speech is pitched. It is about engaging the racial environments in which speech is formed. It is about knowing that black speech is always about much more than what things are said but about how those things are said. And how those things are said involves, of course, the mechanics of grammar, the intonations, the pace, and the flow of black rhetoric, but it includes as well the political and social realities that weigh on the tongue as mightily as the local dialects and accents that rumble in the mouths of citizens.

How black folk are heard makes a big difference in how black folk are perceived in the minds of those with the power to make decisions about their existence. Beliefs about blacks invariably get focused on what they are talking about, and how they are talking about it, and all of that is seen as an index of their intelligence and humanity, their stupidity and savagery. Language means that for other folk too, but just not as intensely, or with as much weight, as for blacks, at least in the United States. The social horizons of black people widen or narrow through words that flow from their mouths; their destinies are shaped by how those words are heard. Still, it is odd, even dispiriting, that the broad public remains largely ignorant of black oral and verbal performance.

Even the most powerful man in the world cannot escape how black speech is heard and read. No matter how high Obama ascends, he is brought back down to the inescapable fact of his blackness and the way he speaks it fluently in white contexts not used to hearing blackness as much as exploiting it. No matter what others think of black language and its rudiments and permutations, in Obama's mouth they have to hear it in their ears, on their televisions and radios, as black speech — as his black speech invades their politics, contaminates their legislative bodies, and is fidgeted over and parsed by the Supreme Court, whether it involves the Affordable Care Act or the Voting Rights Act. Obama's black speech has now become America's way of speaking and of being heard by the world. That may be why, in part, there is such resistance to Obama's policies, because those policies are rooted in black speech: there is a lot of resistance to the uppity character of black speech, no matter how free it is of "Negro dialect," as Senator Harry Reid memorably phrased it.[16]

The far-right Tea Party and the conspiracy theorist birthers despise Obama so much that they want to banish him from Americanness. They want metaphoric sovereignty — well, perhaps they really want the sovereignty of metaphor — over Obama's body: they want to unbirth his existence, uproot him from American soil, foreclose against his house of American identity and offer him a subprime loan of American political capital. The big problem is that Obama has set the terms, symbolically, and sometimes literally, for how America behaves (mind you, that is not a small problem for progressives, who accuse him of rubber-stamping imperialist agendas), and thus they must challenge his legitimacy to act in such an authoritative fashion. Despite the claim of the right wing that it is pro-life, it wants to retroactively abort Obama's existence, purge him from the record as unofficial and illegitimate, remove his legislation from the books, repeal "Obamacare," and wipe the record clean of his political speech. Wiping away his political words also means wiping away his cultural and racial words, the way his body and mouth have left their mark all over America. Obama beat his opponents, and not a few of his ideological allies, politically as well as culturally. He not only licked his opponents with his politics but also licked them with his tongue. The thought is just too ugly for most of them to abide.

Obama must be heard, and understood, in a broader, blacker context, because that blacker context is both in a class by itself and American to the core, as American as Louis Armstrong and Michael Jordan, as American as Condoleezza Rice and Toni Morrison. That blackness is not limiting but freeing; not closed but open; not rigid but fluid. Obama fits along a continuum of black expression and, depending on the circumstance, slides easily from one end to the other, from vernacular to "proper" expression, from formal to informal, from high-tone to gutter-dense, from specifying to signifying in the blink of an "I." Obama's "I" is both black and biracial, both American and international. It is not the beginning of isolation but the start of a new quest for national identity joined to the long pilgrimage of global identity that borrows from centuries of speaking and existing. In the process, a lot of switches are being flipped: codes, styles, media, frames, cultures, and races.

There are echoes in Obama of the rigors and ecstasies of the black speech of Jesse Jackson, Al Sharpton, and of course Martin Luther King Jr., all of which harken to the black church. But the complex signifying, verbal devices, oratorical talents, and rhetorical mastery

taken for granted in the black church, for instance, are largely un-
known outside it. Yet there is a linguistic trace in Obama's speech
that leads straight to the black pulpit.

The Divided Legacy of a Prophet

Obama's "audacity of hope" is a phrase that my late, beloved pastor,
Frederick Sampson, who ministered at Detroit's Tabernacle Mission-
ary Baptist Church, first uttered in a sermon many years ago, and
that renowned Chicago pastor Jeremiah Wright repeated in a sermon
that Obama heard.[17] Long before he was ambushed by right-wing
ideologues and media hacks for his infamous rhetorical demand in a
homily that "God damn America," Wright offered Obama a compelling vision of Christian manhood and enjoyed a national reputation
as a remarkable pulpit orator.

In a 1993 poll conducted by *Ebony* magazine naming the nation's
fifteen greatest black preachers, Wright was second only to legendary pulpiteer the Reverend Gardner Taylor.[18] Wright's preaching
genius derives from a mix of numerous strengths. He is one of the
most intellectually sophisticated and scholarly ministers in the land.
He reads widely and thinks deeply about the pressing religious and
social issues that cram his sermons. He possesses a musical voice
with intonations brilliantly regulated during the course of his sermon delivery, rising and falling as emotion and circumstance dictate.
His diction is flawless and his articulation is precise. Wright's international range of reference reflects his command of several languages.
Wright is not afraid to draw on black vernacular to clarify his prophetic point. He moves effortlessly from the streets to the sanctuary
in illustrating the social sweep of the Christian gospel. And he is
unparalleled in the pulpit in tying black theology to black culture to
embody perfectly his former church's motto: "Unashamedly Black
and Unapologetically Christian." That is the Wright Obama loved
and learned from, and refused to disown when he became a political
liability, at least for a long time.

There is little doubt that Obama heard many other masters of black sacred rhetoric at Chicago's Trinity United Church of Christ during his twenty-year membership there, including Sampson, Charles Adams, James Forbes, Vashti McKenzie, Frederick Haynes, Lance Watson, Rudolph McKissick, and Renita Weems. These ministers attract large followings, drawing thousands to hear them preach with educated zeal and verbal artistry, shaping words to sharpen minds, revive spirits, condemn social injustice, relieve the vulnerable, and uplift the downcast. When Obama opens his mouth, many of these same ideas flow out. But a great deal of the prophetic intensity that burns in small but significant quarters of the black church is lost in translation.

When people yoke Obama and King, on white T-shirts or multicolored wall posters, they are linking the amiable politician to the conciliatory visionary who is exalted in annual King holiday celebrations. Of course it may not be in Obama's best interests to highlight a side of King we would rather brush under the rug: the disturbing, challenging, and revolutionary shadow he cast on the social horizon in his later years. Wright blazed into infamy through fiery Internet excerpts from an old sermon that trumpeted a stunning political reversal—not "God bless America" but "God damn America" for its sins. Except for its blunt language, Wright's denunciation is standard for biblical prophets who say that God will send a nation to hell for disobedience and corruption, a theme right-wing evangelicals have been hammering for years from the opposite ideological direction.

King too said that God would judge America and find it wanting. The night he was murdered, found among King's effects were the notes of a sermon he was to preach the next Sunday: "Why America May Go to Hell."[19] This surely is not the King Obama wishes to be identified with in the American imagination.[20] But it is the King that Obama and others may have to confront if they are to embody and extend the great man's enduring legacy. It is not true that Obama and Wright differ because one buys into King's vision and the other does not; rather each man fixes on a different theme and time in

King's life: for Obama, the optimistic early King, for Wright the revolutionary later King. When Obama speaks, the later King is lost to memory, much the way King's name got lost in his speeches the closer Obama got to the White House.

Before 1965, King was upbeat and bright; he believed white America would change as its conscience was lovingly seared. That is the King Obama, and America, love. After 1965, King was darker and angrier. He did not think America would move forward without considerable coercion. King's skepticism and anger were toned down when he spoke to white America. When he got before black audiences, King's rawer feelings spilled over. His passionate oratory rang out against poverty, war, and racism in hundreds of sanctuaries and meeting halls across black America. King sadly concluded that the Civil Rights Act of 1964 and the 1965 Voting Rights Act "did very little to improve" northern ghettos or to "penetrate the lower depths of Negro deprivation." In seeking to end black poverty, King told his largely black staff in 1966 that blacks "are now making demands that will cost the nation something . . . You're really tampering and getting on dangerous ground because you are messing with folk then." What was King's conclusion? "There must be a better distribution of wealth, and maybe America must move toward a democratic socialism."[21]

Exactly a year before he perished, King shared his opposition to the Vietnam War in front of a largely white audience at Riverside Church in New York. But he reserved perhaps his strongest antiwar language for sermons before black congregations. In his own pulpit at Ebenezer Baptist Church in Atlanta two months before his death, King raged against America's "bitter, colossal contest for supremacy." He argued that God "didn't call America to do what she's doing in the world today," preaching that "we are criminals in that war" and that we "have committed more war crimes almost than any nation in the world." King insisted that God "has a way of saying, as the God of the Old Testament used to say to the Hebrews, 'Don't play with me, Israel. Don't play with me, Babylon. Be still and know that I'm God. And if you don't stop your reckless course, I'll rise up and break the backbone of your power.'"[22] That is a kinder, gentler version of Wright's "God damn America."

Perhaps nothing might surprise — or shock — white Americans more than to discover that King said in 1967, "I am sorry to have to say that the vast majority of white Americans are racist, either consciously or unconsciously."[23] In a sermon to his congregation in 1968, King openly questioned whether blacks should celebrate the nation's 1976 bicentennial. "You know why?" King asked. "Because [the Declaration of Independence] has never had any real meaning in terms of implementation in our lives."[24] In the same year, King bitterly suggested that black folk could not trust America, comparing blacks to the Japanese who had been interned in concentration camps during World War II: "And you know what, a nation that put as many Japanese in a concentration camp as they did in the '40s . . . will put black people in a concentration camp. And I'm not interested in being in any concentration camp. I been on the reservation too long now."[25] Earlier, King had written that America "was born in genocide when it embraced the doctrine that the original American, the Indian, was an inferior race."[26]

Had YouTube been in play when King walked the earth, he might very well have been slammed as an anti-American racist who allowed politics to get the best of his religion, the same way many view Wright today. Some may say that forty years ago King had better reason for bitterness than Wright does in our enlightened "post-racial" America. But that would put too fine a point on arguable gains. It would also prove just how much we do not know about the prophetic energy of the black church.

The black church was born when politics got the best of the white church and kept its leaders from extending Christian love to black folk. Blacks left a church that favored white supremacy over religious kinship to praise God in their own sanctuaries, on their own terms. Courageous slave ministers like Gabriel Prosser, Denmark Vesey, and Nat Turner hatched revolts against slave masters. Harriet Tubman was inspired by her religious belief to lead hundreds of black souls out of slavery. For many blacks, then and now, religion and social rebellion go hand in hand.

For most of their history, the black pulpit has been the freest place for black people; the church has been the place where blacks gathered to enhance social networks, gain education, wage social struggle, and to express the grief and glory of black existence. The preacher was one of the few black figures free from white interests and unbound by white money. Because black folk paid his salary, he could speak his mind and that of his congregation. The preacher often said things that most blacks believed but were afraid to say. He used his eloquence and erudition to defend the vulnerable and assail the powerful. King richly enlivened the prophetic tradition.

Obama has seized on the early King to remind Americans what we can achieve when we allow our imaginations to soar high as we dream big. Wright has taken after the later King, who uttered prophetic truths that are easily caricatured when snatched from their religious and racial contexts. For instance, Wright built to his "God damn America" climax by probing the political roots of racism in his sermon. Wright's homily rages against the oppression of vulnerable black citizens and scorns America for believing it is God, a mistake made long ago by Israel and Babylon. "The government gives [blacks] the drugs, builds bigger prisons, passes a three-strike law and then wants us to sing 'God Bless America.' No, no, no, God damn America, that's in the Bible for killing innocent people. God damn America for treating our citizens as less than human. God damn America for as long as she acts like she is God and she is supreme."[27] An incurable love fueled King's hopefulness and rage and united them in his early and later periods. King's example reminds us that as we dream, we must remember the poor and vulnerable who live a nightmare. And as we strike out in prophetic anger against injustice, love must cushion even our hardest blows.

Obama need not speak in rage to challenge white America to live up to its ideals, just as he challenges black America to do the same.

But he may have calculated that King's mantle was too much to handle. Again, the leader's name noticeably disappeared from Obama's speeches as he got nearer to the Oval Office. Though he may have choked up when rehearsing his acceptance speech for the nomination of the Democratic Party in 2008, timed to coincide with the forty-fifth anniversary of the March on Washington, Obama choked off formal recognition of King in his speech by referring to the leader only as the "young preacher from Georgia." Where his speechwriter may have aspired to a felicitous turn of phrase that captured King's vocation and expressed his race-less universality, Obama might have inserted King's name as a metaphor for rescuing the unnamed from their anonymous fates.

If a man whose name adorns a national holiday can be slighted for fear that he will eclipse the moment or otherwise taint the transcendent aspirations of a campaign that his very blood made possible, it is a slight that should be called out. Obama may have felt that it was enough to set the tone of the night with the civil rights pageantry that preceded his speech, featuring John Lewis and King's children. That does not, however, make up for failing to mention his rhetorical and symbolic benefactor once he arrived at the big dance, especially since he had borrowed King's style and phrases to get there.

The slight was repeated and compounded the day after the celebration of King's eightieth birthday, when his name was not mentioned at all in Obama's first inaugural address. King should have been named, not because he was black, but because, like Lincoln—who was named, not referred to in the speech as a tall man from Illinois with a long beard—King was a great American who forged the path of destiny for a young black man from Hawaii who made Chicago his home before making the White House his castle, and who might not have had the opportunity to make that speech had King not been murdered in Memphis. The distance from King's assassination to Obama's inauguration is a quantum leap of racial progress with a timeline neither cynics nor boosters could have predicted. As Obama continues to mine the rhetorical riches of the prophetic black church and of King's eloquence, he might offer the nation greater gain by sampling as well their courage, and learn to risk his reputation just a bit to tell the truth and serve his nation even better.

Obama's dilemma of being leader of the free world while having sprouted from black religious soil raises the question of how much black sacred rhetoric can survive the flight from prophecy to political power. Much of the rhetoric of the black church has been drawn from the imagery of the oppressed and powerless. Obama said in his memoir *Dreams from My Father* that when he attended Trinity Church in Chicago, he "imagined the stories of ordinary black people merging with the stories of David and Goliath, Moses and Pharaoh."[28] Now he is in the unique position of having reversed the course: from the ranks of the oppressed has come a ruler. Pharaoh has yielded to Moses. David is now Goliath. Does a leader who sees himself as Joshua have the heart to carry the hopes of his people into the Promised Land? I will take that up later, but for now, I want to address how Obama brilliantly, shrewdly, and often disturbingly grappled with Jeremiah Wright's ideas in one of the most famous race speeches in the nation's history.

Wright Turns, Wrong Lanes

During the primary season of 2008 Obama did well on Super Tuesday and chalked up eleven primary victories across the nation in February during Black History Month as the candidate made unprecedented history of his own. But in March, black history gave way to white hysteria when snippets of old sermons by Jeremiah Wright were purchased from Trinity Church and reported on by ABC before quickly migrating to cyberspace, where they were endlessly looped on YouTube. Besides his infamous invocation that "God damn America," Wright argued that America is run by rich whites, that AIDS was invented to destroy black folk, and that our country is "the US of KKK A," even as he mocked Secretary of State Condoleezza Rice as "Cond-amnesia." In the ominous shadow of 9/11,

Wright lit a prophetic fuse by declaring that when America bombed Hiroshima and Nagasaki, "we nuked far more than the thousands in New York and the Pentagon, and we never batted an eye." Wright charged that the United States supported "state terrorism against the Palestinians and black South Africans and now we are indignant, because the stuff we have done overseas is now brought right back into our own front yards." And with a dramatic flourish not uncommon in the theater of black prophecy, Wright twirled his hand above his head to punctuate his deliberately slowed cadence in declaring that "America's chickens are coming . . . home . . . to roost."

Obama bravely attempted to place Wright's comments in a broader racial and social context.[29] He claimed in an interview with the *Chicago Tribune* that Wright was like "your uncle who says things you profoundly disagree with, but he's still your uncle." Obama characterized Wright as an "aging pastor" who had reached maturity in the sixties, and like many black men "of fierce intelligence coming up" in that era, Wright carried "a lot of the language and the memories and the baggage of those times." Obama said he "profoundly" disagreed with Wright's views and that he, Obama, represented "a different generation with just a different set of life experiences," who saw "race relations in just a different set of terms than [Wright] does." Obama acknowledged Wright's religious influence and theological prominence while carefully separating his pastor from his political orbit. Foreshadowing the tack he would soon take in his race speech, Obama refused to disown Wright and voluntarily asked himself a question that was on the minds of millions: "And so the question then for me becomes what's my relationship to that past? You know, I can completely just disown it and say I don't understand it, but I do understand it. I understand the context with which he developed his views, but also can still reject [them] unequivocally."[30]

Obama's laudable effort to clarify Wright's views blended nicely with the shrug offered by most blacks to Wright's comments, knowing as they did that such thoughts routinely boom from the prophet's microphone in the black church. It is true that pastors who address the spiritual and moral needs of their church members by praying for them when they are sick or counseling them through trouble out-number ministers who do all of that and still find time to rail against society's ills. But socially minded pastors often take to pulpits across the nation to object to unjust wars or unfair labor practices or to scold the powers that be for their neglect of the poor and vulnerable.

Even when black Christians do not share their pastors' politics or agree with their slant on scripture to support an argument, they are not alarmed to hear clergy take on the government or powerful public figures. Prophets in the pulpit are like blues artists: they sing dirges that brim with irony and tragedy; they try to lessen suffering through comic gestures of defiance; they amplify, and then diminish, black pain through gut-wrenching poetry; and they trade freely in bombast and hyperbole to beat back the demons of fear and anxiety. Most black folk get the moral intent of black prophecy and believe that they and their divine mouthpieces have a God-given right to ex-press their gripes in the privacy of sacred space. They do not mistake anger at America's imperial excesses for hatred of the nation or a denial of the wonderful changes that can unfold in the country when courage weds imagination.

The Wright debacle proved that many citizens have no idea how instrumental the church has been to black progress. The black church is quite literally a sounding board for vetting ideas and voic-ing frustrations so black folk can stay sane in the midst of America's denial of black humanity. Prophetic preachers also function as exis-tentialist philosophers and absurdist novelists. Like homegrown Jean Paul Sartres and Frantz Fanons, prophets spout theories about what harms or unnerves us; at other times they weave stories about mak-ing the black self into something fiercely beautiful. Their sermons help black folk resist the evil seductions of white supremacy. Very few members go away from such homilies without hope for their fu-ture or belief in their individual importance. Very few leave without a sense of God's care for their burdens and traumas.

But it quickly became clear that the white mainstream had little patience for either Wright's existentialist blues or Obama's dutiful deconstruction of them. When Obama's literary criticism and cultural analysis failed to win the day, he turned to outright condemnation to distance himself from Wright's controversial views.

But Obama would have to help bind the nation's racial wounds by showing familiarity with the ancient injuries of bigotry even as he looked to a future unscarred by race. He would also have to reduce the swelling of trouble in his campaign caused by a combination of devilishly irritating forces: political foes in his own party, right-wing ideologues, an insatiable but poorly informed media, and an insecure public that was very nervous about how Wright's words might have rubbed off on Obama's humanity. Obama had sold himself as a conciliatory figure who could bring all sides together because of his biracial heritage and his unruffled demeanor. Now that promise seemed ruined; his chance for victory appeared slim, and there was grave doubt that he could make the kind of history millions had envisioned for him. If Obama had been searching for the right time to get the monkey of race off his back, he certainly could not have picked a more dramatic moment to cast off the metaphoric long-tailed mammal: he was getting battered in the media and was facing critical primary battles in April. With all that pressure Obama mounted a rostrum in Philadelphia on March 18, 2008, and rode into history with an eloquent speech that saved his campaign and ultimately got him elected president. Obama's oration justly won wide acclaim while laying bare troubling aspects of his reading of the nation's racial compact.

Speech of a Lifetime

Obama stepped onstage at Philadelphia's National Constitution Center in full patriotic splendor with four Stars and Stripes adorning either side of the podium. Like Martin Luther King Jr. forty-five years before him in his "I Have a Dream" speech on the National Mall, Obama strove to link his language to America's glorious if imperfect past. In King's case, the dream metaphor cascaded down the mountain of history that loomed symbolically over his shoulder in the presence of Abraham Lincoln's majestic memorial.[31] The title of Obama's speech, "A More Perfect Union," rooted him in the vast reaches of the country's origins and tied him to the Founding Fathers who once thronged a hall across the street from where Obama now stood to set a nation on its path to democratic adventure. Obama brilliantly condensed and conveyed the sweep of American history: the crafting of the Constitution, the stain of slavery on that document and the national conscience, the fateful specter of the Civil War, and the noble resistance of valiant souls in the civil rights movement. Obama boldly drew a direct line from history to his story to suggest that the nation's future resembled his family past and present—a daring move that teemed with the sort of confidence of destiny that falls every so often on a national figure. His biracial story, Obama concluded, "seared into my genetic makeup the idea that this nation is more than the sum of its parts—that out of many, we are truly one."[32]

Obama briefly described how his message of unity countered traditional lines of division in the South and a media hungry for signs of racial polarization—and suspicions about the candidate being too black or not black enough. Obama acknowledged that racial divisiveness had recently flared at two extremes. On the one hand, former congresswoman Geraldine Ferraro, the Democratic vice presidential candidate in 1984, and a Hillary Clinton backer, although unmentioned here by name, had made the accusation, in Obama's telling, that his candidacy was "an exercise in affirmative action" driven by "wide-eyed liberals" who wanted to "purchase reconciliation on the cheap." Ferraro had complained in March 2008 that Obama was receiving preferential treatment in the campaign because he was black. "If Obama was a white man, he would not be in this position," Ferraro protested. "And if he was a woman of any color, he would not be in this position. He happens to be very lucky to be who he is. And the country is caught up in the concept."[33] It was eerily reminiscent of Ferraro's dig at Jesse Jackson during his '88 presidential run, when she said that because of his radical politics, "if Jesse Jackson were not black, he wouldn't be in the race." Jackson shot back with typical verbal bravura, "Some people are making hysteria while I'm making history."[34] On the other hand, his pastor Jeremiah Wright had used "incendiary language to express views that have the potential not only to widen the racial divide" but that also "denigrate both the greatness and the goodness of our nation; that rightly offend white and black alike."

Obama instinctively knew how much terror Wright's words struck in the heart of whiteness. He borrowed the sociological lens of W. E. B. Du Bois to peer into the souls of white folk. He slipped inside the chambers of their concerns and echoed their disgust at Wright's words. He may have let them know that they were not crazy to feel as they did, but he also courageously told them that Wright was not nearly as crazy as they feared. It was an impossible balancing act, and one that Obama did not pull off perfectly, but the effort to do so proved his genius and his gumption. Obama knew that some whites understood that prophetic preaching was occasionally overstated to match the hugeness of the targets at which it aimed—nothing short of the national way of life that thrived on denying the views of black folk whose mistreatment made them what Malcolm X called "victims of American democracy."[35] But those whites were not the ones whom Obama sought to reassure; rather it was more middle-of-the-road whites who lived in the mythical heartland that Obama had in his sights. Obama surely knew that most blacks were not bothered in the least by Wright's comments, except as they affected Obama's chances to get to the White House. Many blacks and whites agreed with Wright that rich whites run America. All forty-three of the U.S. presidents before Obama were white. Most CEOs of Fortune 500 companies have been white. And most members of Congress have been white. Throughout history, white Americans have been dominant and nearly unchallenged in every quarter of power that matters.

Lots of black folk and many others—including former diplomat and Reagan appointee Edward Peck—agreed with Wright that
America's vicious misdeeds around the world had in part come
home to haunt the nation in the events of 9/11. Like Peck, Wright
was not celebrating the attack but using it to analyze our situation
and warn us against future folly. In short, America's suffering at the
hands of terrorists was not praiseworthy but predictable. Wright
noted in his sermon "The Day of Jerusalem's Fall," delivered a year
after 9/11, "I heard Ambassador Peck on an interview yesterday . . .
[point] out that what Malcolm X said when he was silenced by Elijah
Muhammad was in fact true—he said American chickens are coming
home to roost." Wright offered his prophetic diagnosis with nearly
the same phrase that journalist Mike Wallace used when he reported
on the Nation of Islam and Malcolm X in the sixties in a documentary he called *The Hate That Hate Produced:* "Violence begets violence.
Hatred begets hatred. And terrorism begets terrorism. A white ambassador said that y'all, not a black militant. Not a reverend who
preaches about racism. An ambassador whose eyes are wide open
and who is trying to get us to wake up and move away from this dangerous precipice upon which we are now poised. The ambassador
said the people we have wounded don't have the military capability
we have. But they do have individuals who are willing to die and take
thousands with them. And we need to come to grips with that."[36]

Wright shrewdly cited a career diplomat with impeccable conservative and patriotic credentials to support his argument that
America's violence abroad had boomeranged in domestic disaster.
Obama surely could not agree with Wright's analysis and remain
a viable presidential candidate; but he might have helped explain
Wright better by placing him in an honored tradition of prophets
who denounce America's sins rather than dismissing him as incendiary and divisive. Inflammatory speech is not the litmus test for
morality; after all, Jesus argued that he was coming to divide mother
from daughter and to bring not peace, but a sword to all those who
sought refuge from an unjust political order. And while many black

folk do not agree with Wright that AIDS was invented to kill black folk, they do believe that science has been misused to harm black life. This fear was tragically confirmed in the Tuskegee experiment in which scientists tested the effects of syphilis on nearly four hundred unknowing black subjects between 1932 and 1972 and failed to offer them penicillin when it became clear in the 1940s that the new drug could cure the disease.[37]

The positive take on Wright's ideas among black folk did not mask their hostility to any force that stood in the way of Obama's run for the White House. Talk show host and social activist Tavis Smiley got the bitter brush-off from millions of blacks who soured on his rigorous questioning of Obama's political aims and his failure to support a black agenda.[38] The history of black America's quest for the presidential Holy Grail had come down to whatever chances Obama had for securing the Democratic nomination. Obama had easily become the biggest thing since King in black America, and the prospects for his success, and the barriers to his ascent, even if erected by well-meaning figures, were taken with deadly seriousness.

Prophets and Politicians

The conflict between Wright and Obama was a showdown of two archetypes in black America: the prophet and the politician.[39] If history is our guide, the prophet usually prevails in any skirmishes between the two because the prophetic tradition has given black America its biggest boost through brave figures like Absalom Jones, Richard Allen, Denmark Vesey, Nat Turner, Sojourner Truth, Frederick Douglass, Harriet Tubman, Henry Highland Garnet, Reverdy C. Ransom, W. E. B. Du Bois, Paul Robeson, Nannie Helen Burroughs, Septima Clark, Ella Baker, Fannie Lou Hamer, Martin Luther King Jr., Joseph Lowery, Benjamin Hooks, Jesse Jackson, and Al Sharpton. Most of these figures were clergy or were inspired by the black church to fight injustice. The prophetic voice spoke black politics into existence and cleared space in the American landscape for the just exercise of black power by politicians like Hiram Revels, Blanche Bruce, Robert Smalls, Oscar De Priest, Arthur Mitchell, William Dawson, Charles Diggs, John Conyers, Shirley Chisholm, Charles Rangel, Richard Hatcher, Kenneth Gibson, Barbara Jordan, Coleman Young, Harold Washington, James Clyburn, Maxine Waters, Barbara Lee, Carolyn Cheeks Kilpatrick, Elijah Cummings, Kendrick Meek, Marcia Fudge, Joyce Beatty, and a host of others. These political figures borrowed energy from the prophets and depended on their constant harangue to relieve them of that role, at least as they negotiated their way in the white establishment. Ordained ministers and congressmen like Adam Clayton Powell Jr., Andrew Young, John Lewis, Walter Fauntroy, William Gray, Floyd Flake, and Emanuel Cleaver blended the political and the prophetic and brought religious style to politics and political swagger to ministry.

Black folk understand that even though prophets and politicians sometimes share professional instincts and hunches, they have distinct roles to play. Most black politicians know that they owe their origins and influence to a sacred sphere that they dare not insult or ignore. Prophets are the parents of the politician and the rightful guardians of the black community's spiritual and moral trust. Prophets also recognize that elected officials embody the authority and legitimacy for which the black community yearns. Prophets are zealous to bring moral and social correction by urging the sort of radical change most political figures are afraid to endorse. Politicians do not usually experience the free speech the prophet enjoys, and yet, for that reason, they move more easily within the corridors of power to argue for benefits that are the secular version of spiritual blessings. Yet when prophets fall into dispute with politicians, they often exploit their religious authority to snag the allegiance and love of the folk.

Barack Obama's rise may not be the first instance of the prophetic taking a backseat to the political, but it is surely the most visible example in black history. That is because Obama's emergence as the nation's first black president and the world's most powerful man—and the most powerful black figure in global history—has been viewed by arguably more blacks than ever before as divinely inspired. The political easily trumped the prophetic in Obama's case because his candidacy co-opted one of social prophecy's crucial aims: the embodiment of power in a figure who looks like blacks and who will justly represent their interests. But that aim may have been retailed in the black rush to rule: a black Realpolitik drove blacks to baptize the need for viable representation in the murky waters of political symbolism. What came out in the wash was the understandable if unsatisfying desire to have a black face in the highest possible place as the exhaustive and definitive symbol of the black political journey.[40] Such a view threatens to reduce the tradition of black politics to a single personality—a charge, by the way, often made against so-called messianic forms of black leadership, like those glimpsed in King, Jackson, and Sharpton, which rely on charisma more than institutional power. Obama's example was supposed to reverse this trend, and yet only King has arguably been a bigger messianic figure in black circles, even if Obama is a crossover messiah whose exclusive ties to black piety or politics are dramatically snipped. Unlike King, Jackson, and Sharpton, Obama is not a black leader but a leader who is black, though such a distinction rings hollow in millions of black chests that swell with pride in Obama's extraordinary accomplishments.

For those blacks, only Martin Luther King Jr. exceeds Obama's place in black history, and that priority may be temporary as Obama's reign as the face of America eclipses for many even King's exalted status. The stenciling of Obama's face next to King's on T-shirts symbolizes the passing of the torch of "most important black man in history" from *the* prophet to *the* politician.

That backdrop made it clear that while black folk understood why and how Wright's controversial words were spoken, they also believed that Obama had to distance himself astutely from his former pastor. After all, whether he or the rest of blacks like it or not, Obama represents not simply himself but all black folk and all their traditions, with all their inherent conflicts, all at once. It is a sign of how confused racial politics are that, at the same time, Obama is seen by many whites as quite unlike most other blacks, and therefore an exceptional member of the race. In the rift between Wright and Obama, the political leaped onstage and seized the historical microphone as the prophetic watched from the wings at a silent distance. Obama's and Wright's eventual war of words underscored a paradox in black America: the prophetic was widely seen as an obstacle to the political power it has always argued should exist. Wright's prophetic words were viewed as a roadblock to Obama's divinely appointed political destiny.

If God could not be said to be only on Obama's side, God was at least squarely in his corner as Obama struggled to honor Wright as much as possible by uplifting his priestly and pastoral roles while criticizing his prophetic pose. Obama shrewdly split the difference between complementary aspects of Wright's religious vocation — actually he divorced them outright and set them at war to pursue a remarkable if risky undertaking: to satisfy blacks *and* whites with the critical embrace of his pastor. Obama had to convince blacks that he was not really dissing Wright in his pastoral role, a taboo in black circles where "mother" and "pastor" are keywords of reverence and portals of heavenly blessing. Besides, the bread and butter of pastors is to deepen the congregation's spiritual life, while all the political stuff is extra trimming. If Obama pummeled Wright's politics but praised his priestly duties, he would come out of the skirmish unscathed in most black quarters.

That was only half the battle, though, and not necessarily the more important one for his immediate purposes. Obama also had to convince whites that he detested Wright's racial and political perspective, which had little to do with either Wright's primary religious duties or with Obama's political views. "He hasn't been my political adviser," Obama told the media. "He's been my pastor."[41] Obama was surely no prophet, but he appeared to be playing one on television to great effect. Obama's extraordinary visibility and far bigger bully pulpit offered him an unfair advantage over Wright in their competing attempts to shape the perceptions of the righteous duties of the black preacher and to shed light on the prophet's real role. Though Obama's was by far the most powerful interpretation of prophetic power, it was not the only one going—in fact, it surely was not the one favored by critics familiar with prophecy's rich history. It also contrasted sharply with the interpretation offered by Wright when he broke his silence and sought to defend himself in the media. But it was certainly the one favored by the masses of black folk who fervently prayed for Obama.

When Obama said in his "More Perfect Union" address that Wright used "incendiary language to express views that have the potential not only to widen the racial divide" but also "denigrate both the greatness and the goodness of our nation; that rightly offend white and black alike," he snatched the mantle of the prophet, and the prophet's benefits too, one of which is that one need not tell the literal truth to get at the moral truth behind one's words. Obama thus exploited his honorary prophetic standing to skirt Wright's intention and instead divine the effect of the prophet's words on the mainstream's eardrums and worldviews. Obama also took license to do all that he could ethically do to secure the nomination and fulfill his divine destiny. Obama did not have to believe his run for the White House was divinely inspired for the idea to catch on. All he had to do was to blend his political aspirations with the aspirations of black folk and the nation and let the critics figure out what it all meant.

Forgiveness Knowledge and Tough Choices

Black folk were even willing to forgive the thumbs-down Obama gave to Wright's words in an effort to appease whites who resented the criticism of their culture by unapologetic black voices. We've seen that Wright's words were meant in truth not to denigrate the greatness and goodness of America but to help the nation recognize that it was hurtling away from democracy. Wright's words were not offensive in the least to those who understood the nature of prophetic speech throughout the nation's history. Damning sinners to hell is a raging prophet's heaven, whether it was Jonathan Edwards in the eighteenth century or Billy Graham in the twentieth. Even earlier, Puritan divines like John Winthrop preached a form of sermon called the jeremiad to warn folk of the divine retribution that lay in store should they neglect doing justice in the land.

Black folk were eager to extend Obama the privilege of using what might be termed *forgiveness knowledge:* the pardon offered in advance to one who gives a negative spin to ideas that might otherwise be interpreted positively if the choice to do so were not deemed so costly to the greater good. Forgiveness knowledge preempts criticism of a figure who says harsh things about a person or beliefs because that figure is unfairly painted into a corner with no option except to act and speak as she does. Forgiveness knowledge is knowledge that, once one has it, one will be forgiven for having to possess and use because the conditions do not allow for a reasonable or effective alternative. Forgiveness knowledge in this case is best summarized in the question many blacks asked: "Well, what other choice did Obama have than to say what he said about Wright under the circumstances?" Blacks forgave Obama even before he spoke about Wright because they believed that he had no choice but to agree with the white mainstream's negative reading of Wright's words as a starting point before attempting to set those words in historical context. Black folk forgave Obama because they realized that if he had not offered his criticism of Wright first, he would not have had the opportunity to offer his qualified defense of him later.[42]

Obama's offering at best a qualified defense of Wright was a choice seen as largely out of his hands. But the terms of Obama's defense rested solely in his imagination. Obama's forgiveness knowledge would permit him to say why he loved Wright and loathed his words in one fell rhetorical swoop. True, he would have to misrepresent Wright a bit so that he might represent him as best he could to a white mainstream that simply was not accustomed to hearing complex black views about too many issues. The fault was not entirely, or even primarily, Obama's but that of a white mainstream that had the power to dismiss what it did not want to know as illegitimate and immoral—or in the case of Wright, as utterly illogical and indefensible. Obama had to, in effect, tell a white lie to a white audience to get some black truth through. In this case one must heed the street admonition vigorously: do not hate the player but the game. Obama was not making up the rules but following them to perfection. Obama told as much truth as white ears can hear about black life. He also gave whites permission to be angry with Wright without fearing that they would be seen as bigots, and without feeling that they alone were responsible for racial healing.

Obama did not feel he could be nearly as generous to black folk—or that it was even necessary to try. He relegated black anger to the sixties, as if no current troubles might incense a righteous and reasonable black person. Obama also put black anger at racism in the same moral orbit as white resentment of having to bother with fixing the racial problem. And he may have overplayed the extent of blacks' displaying a victim mentality while underplaying their true victimization.

Obama admits that he knew Wright was an "occasionally fierce critic of American domestic and foreign policy" and that he made controversial statements with which Obama strongly disagreed—just as the candidate's listeners sometimes disagreed with their pastors, priests, and rabbis. (Under current racial conditions, Obama's qualified defense of Wright here seems all but heroic.) Wright is taken to task not simply for being controversial and divisive but for "a profoundly distorted view of this country—a view that sees white racism as endemic, and that elevates what is wrong with America above all that we know is right with America; a view that sees the conflicts in the Middle East as rooted primarily in the actions of stalwart allies like Israel, instead of emanating from the perverse and hateful ideologies of radical Islam."

Obama may have carelessly accused Wright of believing that white racism is endemic and peculiar to the nation without realizing he had already made a similar claim in his speech. Obama had just called slavery America's original sin, which sounds fairly endemic as well, since it is the notion in Christian theology that Adam's original act of disobedience causes all human beings to be born in sin. White racism is the womb in which slavery took shape, or to shift metaphors, it is the moral ecology, the spoiled Garden of Eden, as it were, in which slavery took root and makes sense. Judging by how they approach their respective careers, Obama and Wright may not be as far apart as advertised in their method of combating racism. Most good preachers have battled the notion that it is impossible to fight sin, a sentiment that flares in the fatalistic cry "The devil made me do it." And most politicians have mounted campaigns against the sort of cynicism that is heard in the jaded lament that "all politicians are corrupt." For prophet and politician alike, the remedy to our problems is conversion—a transformation of attitude and belief as each figure touts a change we might believe in. Some conversions are cool and mellow, while others are sparked from the threat of fire and brimstone. The style of conversion might differ but the goal is still the same: to change the hearts and minds of human beings so that they become believers and remain faithful to the cause. No matter how much Obama and Wright differed as they sought to cast out the demons of racial bigotry, they were both motivated by the same desire to see change come.

Obama may also have undersold his former pastor's critical patriotism. Wright's vigorous disputes with white racism express a robust belief in the country's ability to change. Why would he try to get things right by talking about how bad things are if he did not believe they could be changed for the better? Like all good prophets, Wright had to tell the bad news before he got to the good news of the gospel. He had to say what is wrong with America before he could say what is right with the nation. Politicians operate the same way on a less apocalyptic scale when they cast doubt on an opponent's record:

it is all mess and madness before the change they envision takes hold—or at least that is what they would have us believe. Prophets and politicians are both heavily invested in painting the status quo as sorry and hopeless until their respective solutions help set things right—in the case of the prophet, his brand of preaching, and in the case of the politician, her brand of politics.

Obama reinforced the sense that Wright was wrong about race by suggesting he was wrong about foreign policy, a clever if cynical rhetorical gesture that may not withstand scrutiny, since one has no necessary or logical relation to the other. Obama took issue with Wright's reading of Israel's aggression against Palestinians, though not enough to keep Obama from saying a week before his race speech at a small fund-raiser in Iowa, "Nobody is suffering more than the Palestinian people."[43] It is clear that one can love Israel and support its security against radical Islamic attacks and yet, like thousands of Jews around the world, be critical of Israel's policies against Palestinians. Obama's criticism may have strengthened the perception that his former pastor was unreasonable and dangerous. Obama was also curiously silent about his pastor's take on acts of American terror and aggression against Native Americans, Grenadians, Libyans, Iraqis, Japanese, and South Africans, as if such acts do not cause the same offense or concern for people of color, as if they do not really matter in the same way.

Obama argued that Wright's racially charged comments distracted the nation from solving "monumental problems—two wars, a terrorist threat, a failing economy, a chronic health-care crisis, and potentially devastating climate change; problems that are neither black or white or Latino or Asian, but rather problems that confront us all." But race is a problem that confronts us all too; the proof is that nothing Obama said about these other issues, and nothing anybody associated with him said about these issues, threatened to derail his candidacy quite like Wright's remarks, which touched the deepest and most sensitive nerve in the nation. Race is still the country's most volatile issue and its greatest unhealed wound. Race and democracy grew together in the American imagination and were woven into its laws and braided into its beliefs and behaviors. Even Obama's laundry list of crises is not immune to the shadow of race: if one was black or brown, the bad economy had been even worse, and the lack of health care had hit black and brown folk especially hard; as we saw, the initial debate over universal coverage sent the president and his allies running for cover from the bitter racial fallout over his proposed reforms.

Obama later let on in his speech that race is a big problem, calling it "an issue that I believe this nation cannot afford to ignore right now." So it was not confronting race that was the problem with Wright's comments; it was his divisive manner and his distorted philosophy of racism's bleak persistence that were troubling. And yet Obama seemed to agree with Wright when he said we "need to remind ourselves that so many of the disparities that exist in the African-American community today can be directly traced to inequalities passed on from an earlier generation that suffered under the brutal legacy of slavery and Jim Crow." To his credit Obama named some of them: segregated schools; legalized discrimination that kept black folk from owning property; barriers to getting loans to start businesses, securing mortgages, and landing jobs in police or fire departments; a vicious wealth gap and brutal poverty whose origins trace back to American apartheid; and the blight of economic inequality in black families, which is compounded by the lack of basic services in urban black communities. Though lacking the prophetic passion that streams through Wright's dramatic blues complaints, Obama's pointed analysis shows that he and Wright are not very far apart at all in their estimation of the structural forces that sweep over black America and diminish its stability and prosperity. Of course, the ways they emphasize its results surely vary and are dictated by their distinct vocations of prophet and politician. One hollers. One whispers. It is hard not to conclude that if Wright's views of the matter are distorted, then so are Obama's. Yet that hardly kept Obama from drawing distinctions between himself and his former pastor.

The Politics of Anger

Obama attempted to explain Wright by exposing a generational fault line beneath the ground of black existence: older blacks like Wright who came of age in the sixties are engulfed in bitterness and anger that are usually muted in white America but regularly flash at barbershops, kitchen tables, and black churches—a palpable force that is often, said Obama, "exploited by politicians to gin up votes along racial lines, or to make up for a politician's own failings." The implication, of course, is that Obama is a new black whose racial politics are less heated and will not lead to divisive talk or appeals to race to make a point or win an election. The fact that such an approach could be seen as an effective way to untie the racial knot proves we lack sophistication in grappling with race, and that for their own benefit our politicians sometimes pretend that things on their watch are not as bad as they seem, while the nation continues to suffer.

Obama aimed to get rid of one sort of racial difference while reinforcing another: he reassured white America that he shared neither the age nor the rage of blacks who witnessed the flight of Jim Crow up close. That nimble racial dance was brought on by the need to explain black anger without endorsing it; Obama would never get to the White House except as a visitor if he came off as too angry—or for that matter if he appeared angry at all. Obama's cool demeanor took on political significance and boosted him beyond black leaders known to direct passion against prejudice. It may be that those who got many of the benefits that were fought for in the sixties can afford to be more hopeful, perhaps even cooler, while those who have been left behind, and the folk who speak for them, have more reason to complain about what did not get done.

The need to prove that he was not an angry black man led Obama to deny the political usefulness of black anger and to draw questionable parallels between black and white anger, a move that curiously denied white anger's political destructiveness. Black anger has surely been exploited and destructive, but it has also been productive and redemptive. If black folks' humiliation has led to anger—and to doubt and fear—that anger also led to tremendous change. Malcolm X was palpably angry at the horrible conditions of black life and worked to translate his rage into the redemption of black communities. Martin Luther King Jr., too, poured a prophet's rage on the war in Vietnam, and on poverty and racial inequality near the end of his brief but remarkable run in public life.

Obama contended that black anger distracts black folk from "solving real problems," from "squarely facing our own complicity in our condition, and prevents the African-American community from forging the alliances it needs to bring about real change." Obama may have been hinting at a black anger that is more cathartic than catalytic, one that is grounded in venting before it takes off on the wings of aimless speech or hurtful action with no real chance of changing the things that make us angry. The flip side is that black folk who vent their anger in barbershops, kitchens, and churches are not as likely to spill their blood or the blood of others in the streets. And when Obama earlier claimed that the remarkable thing was not "how many failed in the face of discrimination, but rather how many men and women overcame the odds" to "make a way out for those like me who would come after them," it must be remembered that some of those men and women got angry, and some even bore the cross, so Obama could cross over turbulent political waters.

Obama argued in his speech that black anger kept black folk from wrestling with how they have contributed to their own problems. That argument reflects a healthy tradition of blacks taking other blacks to task for not stepping up to handle their own racial business and not policing their own moral boundaries. The ideological pedigree of those who have made such arguments runs from conservatives who look to Booker T. Washington, to black integrationists who admire Mary McLeod Bethune and King, and black nationalists who hold Marcus Garvey, Elijah Muhammad, and Malcolm X in high regard, or who extol Louis Farrakhan today. Anger and action need not be opposed as black folk grapple with their plight. And neither does it mean that black people must always forge primary alliances with folk outside their race to uplift black people; the virtue of certain black self-help traditions is that black folk take the measure and pulse of their community to enable their own flourishing. It is hard to have it both ways: one cannot scold black people for failing to forge alliances with whites who have proven to be skeptical about such alliances, and then fail to applaud blacks who take their destinies into their own hands by working with one another to improve their condition and to relieve the suffering they have created at their own hands.[44]

Obama bravely took on white anger, too, but in a far less forceful manner. He dodged wrestling with white backlash to perceived black progress. White anger has consistently burned in pockets of the culture; it has led to black psychic injury and to the exaggeration of the black threat to white survival as a way to rationalize the hatred of black folk. There is a feeling among many whites that black folk are getting more than they deserve and that it comes at the expense of whites who have been disadvantaged by the push for black rights. Obama made oblique reference to such a feeling when he said that "all Americans" must "realize that your dreams do not have to come at the expense of my dreams." But that general prescription lacks the bite and force of a specific remedy for white anger like the one he offered blacks. Obama's treatment of white anger was far more sympathetic than his grappling with black anger. He paralleled black and white anger; Obama may have seen that as a necessary rhetorical gesture in the effort to make white Americans understand Wright's wrath, but it was a parallel fraught with peril.

Obama had already made brilliant use of parallel in his speech when he argued that we could "dismiss Reverend Wright as a crank or a demagogue, just as some have dismissed Geraldine Ferraro, in the aftermath of her recent statements, as harboring some deep-seated racial bias." In this example Obama, with elegance and economy, drew an implicit parallel between a hated black figure, Wright, and a far less reviled white one, Ferraro. Obama suggested a racial equivalence that had the advantage of taking the conversation forward by inviting whites to see how Wright and Ferraro could be read in similar fashion. The implication was clear: if whites and other critics were going to beat up on Wright, they had to do the same to Ferraro. And by implying that such a charge was not true of Ferraro—though without explicitly stating it, leaving himself a bit of wiggle room to either embrace or deny the claim later, since his campaign had demanded that Hillary Clinton repudiate Ferraro's remarks—Obama made it plain that the charges against Wright were not necessarily true either. It was a nifty logical and rhetorical gesture that might have served Obama well had he applied it to his views of the difference between white and black anger.

False Equivalences

Even this case reveals how Obama's rhetorical tack denies to blacks the advantage of the moral equivalence he draws between white and black anger: he calls Wright's words divisive but refuses to do the same for Ferraro, though her divisive and racially charged comments were directly related to Obama and the presidential race in a way that was not true of Wright's sermons. It is another subtle but suggestive way in which Obama seems to feel he cannot hold white folk's feet to the fire even as he warms up to criticizing black folk explicitly. Normal appeals to "inside group" argument, in which one may say negative things about one's own group that outsiders cannot say, are pretty much a wash, since Obama explicitly rejected racial solidarity as the basis for his campaign, or as a means of getting blacks to vote for him. But the peril of even gently criticizing Obama among blacks proves that what works in theory may not always work out in practice—though it surely works to Obama's advantage, since black folk have largely failed to call him on his flaws. Obama gets the best of the bargain: he does not have to appeal overtly to racial solidarity to be its beneficiary.

Obama further empathized with white resentment when he

noted that most working-class and middle-class whites "don't feel that they have been particularly privileged by their race," that as immigrants they feel that "no one's handed them anything, they've built from scratch" and had "worked hard all their lives, many times only to see their jobs shipped overseas or their pension dumped after a lifetime of labor." These whites are full of anxiety about the future as they witness their dreams disappear, especially in "an era of stagnant wages and global competition," when "opportunity comes to be seen as a zero sum game, in which your dreams come at my expense." In Obama's reading, black progress breeds white anger: when whites "are told to bus their children to a school across town; when they hear that an African-American is getting an advantage in landing a good job or a spot in a good college because of an injustice that they themselves never committed; when they're told that their fears about crime in urban neighborhoods are somehow prejudiced, resentment builds over time."

Obama fails to point out to working-class and middle-class whites who feel that they have worked hard for what they got, and that they do not benefit from their whiteness, that blacks and other minorities feel the same way but with a twist: they too worked hard, built from scratch, saw their jobs sent overseas, had their pensions cut, suffered depressed wages, and endured deferred dreams, but got denied the advantages that even white immigrants could take for granted. In short, black folk are the Ginger Rogers to white folks' Fred Astaire: as the late Ann Richards quipped in paraphrasing a 1982 *Frank and Ernest* cartoon, "Ginger Rogers did everything Fred Astaire did, except backwards and in heels."[45] Black folk have endured everything that white immigrants have endured, except they worked in chains and under the fateful wings of Jim Crow. Unlike immigrants who got the benefit of their whiteness to work from scratch and earn their way into the American mainstream, black folk who were brought or born here often got the short end of the employment and educational stick in being unjustly denied the fruits of their labor.

As Martin Luther King Jr. noted, at "the very time that America refused to give the Negro any land, through an act of Congress our government was giving away millions of acres of land in the West and the Midwest, which meant it was willing to undergird its white peasants from Europe with an economic floor." King preached against those who demanded self-reliance from black people while overlooking the educational, agricultural, and other subsidies offered to white immigrants. "But not only did they give them land, they built land grant colleges with government money to teach them how to farm. Not only that, they provided county agents to further their expertise in learning. Not only that, they provided low interest rates in order that they could mechanize their farms."[46] Not only did white privilege and government assistance enhance the standing of whites, but it also sharply contrasted to, and reinforced, black suffering. That history—and the persistent gap between what even recent white immigrants can expect from our society in comparison to what blacks often receive—demonstrates that hard work does not distinguish white immigrants from blacks, and suggests that those immigrants have not been economically dislodged by the relatively limited efforts at black social compensation.

Obama could not speak plainly for fear of alienating white voters, but the truth remains: what it means to be white in America is to take advantage of opportunities that rest on clear injustice. Unfair laws allowed whites to own property, accumulate wealth, attend schools, and find employment, all of which were systematically denied blacks. Obama had already sketched just how bad the gap between white and black was, yet he stopped short of drawing the conclusion such facts invite: that white anger at the recent benefits offered blacks for centuries of injustice is in truth crying over crumbs from the larger table of white privilege. Black advantage, such as it exists, does not come at the expense of poor and working-class whites, as many believe, or even at the expense of middle-class whites; instead it is the long-delayed gesture of opportunity that should never have been denied to begin with, and that whites were able to hog for themselves. Being told in the sandbox of life that they

now have to share their toys makes many whites angry and resent-
ful—but their rage is not necessarily righteous.

Even poor and working-class whites were not shafted, and still are
not, because of black progress or affirmative action, but because of
the inequalities reinforced by the white elites that, Wright argues,
control American society. That is why it was painful to observe poor
and working-class white folk take to the streets in Tea Party protests
to decry as socialist the relatively tame Obama administration policies that seek to restore to working-class people some measure of
class and economic dignity. Conservative interests with deep pockets
helped to orchestrate such protests, exploiting the class rage of poor
and working-class whites. The privilege of whiteness is not simply
about an economic payoff that can be documented; it shows, too,
when whites are not harassed in stores by clerks who believe they do
not have money to buy goods. It also registers when whites are not
harassed by cops in their homes or in the streets, and when they are
not murdered by police as routinely as black and brown folk. That is
a form of white privilege that extends to all classes of whites; such
privilege can mean the difference between living and dying. Living
when you might otherwise be dead if you were black or brown is the
ultimate form of white privilege.

Obama had to walk a perilous tightrope in talking to whites. Trying to uphold his appeal as a biracial healer and resist being tagged
as the "black candidate"—not to mention putting on the political asbestos suit to handle his flame-throwing pastor—meant that Obama
would not be telling a compelling story about white anger that might
edge nearer to the truth than the story he was sticking with. All that
he said was true, but it was not all that is true and helpful to be said
to explain the origins of contemporary white anger. Unlike what he
did with black anger, Obama failed to offer a brief history of white
fury and one of its central tenets dating back to slavery: the belief
that blacks and other minorities are less than human and unworthy
of solid social standing or equal rights. White anger helped to extend the reign of white supremacy well into the twentieth century.

It is clear that Obama could barely touch on such matters at such a touchy time in the nation, when whites who did not trust him, and who were angry at Wright, might jump ship or never sign on at all. Those circumstances roped Obama into making an awkward mismatch between black anger from the sixties and white anger today.

Black anger for Obama is counterproductive when it extends to the present; white anger, by comparison, is explained *only* in contemporary terms and thus severed from its racist roots. The disturbing result is that it is okay to be white and angry; it is not okay to be black and angry. Obama passes along to blacks the limitations that whites have imposed on him. When Obama quarantines black anger to the sixties, he gives the impression that black folk today are not righteously angry about police brutality, racial profiling, a subprime mortgage scandal that unjustly bled black wealth, the over-incarceration of black folk, and a host of other ills that ravage black life now. By restricting legitimate black anger to the past, Obama avoids grappling with legitimate forms of new black anger and how they might be related to society's and government's failure to treat black folk justly. Obama makes things worse when he dismisses black anger as a dysfunctional generational trait for black America even as he finds contemporary white anger compelling.

It is not that Obama was oblivious to black rage and what we must do to address it. Obama insisted that whites must do their part to forge a more perfect union by "acknowledging that what ails the African-American community does not just exist in the minds of black people," because the "legacy of discrimination—and current incidents of discrimination, while less overt than in the past—are real and must be addressed" in word and deed. For Obama, those deeds include investing in schools and communities, enforcing civil rights laws, making the criminal justice system fair, and providing greater opportunity to the younger black generation; they include adopting the attitude that "your dreams do not have to come at the expense of my dreams," and that "investing in the health, welfare, and education of black and brown and white children will ultimately help all of America prosper." Obama also admitted that white anger has not always been productive, that it has often been exploited by politicians and talk show hosts, and, like black anger, it has not always spoken in polite company.

Obama, however, offered whites a pass when he argued that their resentments should not be mislabeled as "misguided or even racist, without recognizing they are grounded in legitimate concerns," for "this too widens the racial divide, and blocks the path to understanding." Yet Obama refused to say that white anger has distracted whites from solving real problems; that it has kept them from facing up to their own role in their own troubles. Obama dared not suggest that white anger leads some whites to scapegoat black folk, an easier choice than confronting corporate interests and the vicious practices of capital that undermine white people's lives more than the paltry payoff of affirmative action to black folk and other minorities, including white women. Obama did not hint to white folk that their prejudice has cost them too, and contributes to their own suffering because it keeps them from forging ties to people of color and forming a coalition of conscience that might put an end to the economic bleeding they endure.

Role Play

Of course, Obama could not say much, or any, of that because, as Wright noted, he and Obama occupy different universes of social purpose and orbit different spheres of political interest: "I do what pastors do. He does what politicians do."[47] Wright made those comments at the National Press Club on April 28, 2008, where he had gone to cap a round of highly visible public appearances after he had been pilloried in the press and after Obama had spoken about him at length in his race speech and in several press conferences. Wright had already made a successful stop on Bill Moyers's PBS television program, where he had ably defended himself from the unfounded charges tossed his way.[48] After Moyers's show, Wright appeared again on national television in a speech to the Detroit branch of the NAACP that proved he was very smart, droll, verbally gifted, and charming.[49] Wright's Press Club comments about the pastor's and the politician's competing allegiances and contrasting vocational approaches, however, left Obama ticked. He felt that Wright's division of vocational labor was snide and dismissive and painted him as just another huckstering politician who lacked principle while tickling the fancies of the public in order to win office. It was only then that Obama totally rejected Wright and his message and severed all ties with his former pastor.

True to form, Wright was simply distinguishing the role of the prophet from that of the politician. Some black ministers balked when Wright said that God had damned the nation. When Wright was quizzed at the Press Club about Obama's churchgoing habits, many of the same ministers winced when he snapped at the moderator: "He goes to church about as much as you do. What did your pastor preach on last week? You don't know? OK." Yet other black clergy refused to condemn him. Most black folk seemed to draw a practical distinction: Wright surely could not be taken to task for past pronouncements that reflected his prophetic pedigree. The man was simply doing his job, which was to defend black folk against unjust forces like the best black prophets have always done.

But the present was another matter altogether. Far fewer blacks seemed willing to extend to Wright the privilege of derailing Obama's presidential campaign with his battle against false interpretations, misleading characterizations, and relentless media pummeling. Many black folk wanted Wright to take one for the race. Black folk had seen plenty of the prophetic, but none of the presidential, in black America, and they expected Wright to understand that and go gentle into the night of racial discretion, black solidarity, and cultural quiet. Wright's persistent attempt to be heard after he had been savaged in the press was completely understandable; after all, his entire career of brilliantly preaching and faithfully carrying out the duties of a pastor had been reduced to a series of thirty-second sound bites that truncated his vocation with violent precision. Moreover, prophets rarely show up when it is convenient for those in power; enough black folk followed the prophetic script to know better than to think otherwise. They knew that prophets will be prophets.

That seems an acceptable stance when times are normal — meaning during times when no black person has a shot in hell to make the Lincoln Bedroom his own. But these were not normal times; this was not *chronos*, that is, calendar or clock time, but *kairos*, which for Christians means the right time and the opportune moment to fulfill God's purpose.[50] Prophets usually have *kairos* on speed dial; they usually have the right time on theological lock. Since this appeared to be one of those rare moments in black life when the political usurped the prophetic, Wright was unfairly made the poster child for reverse *kairos* — when the prophetic seemed to be the impediment, and not the instrument, of divine delivery. Wright found himself on the wrong side of the black history he had spent his life defending. Without being given a fair chance for rebuttal, Wright lost the battle for black sacred truth that Obama and his legion of followers had at least temporarily won.

Wright was thrust into an untenable position: if he dared to defend himself against the merciless attacks that were being launched his way, he would be seen as hopelessly self-serving and ego-driven, a curious charge since most political figures are nearly forced to exalt themselves as saviors and to tell us that they will not get to save us unless they get elected. But if Wright allowed his name to be taken in vain, without offering a defense, then it might look as if he had no interest in defending the prophetic stream in which he regularly swam. How Wright had come to national prominence only made things worse: he got his brightest shine, and his darkest shaming, because of his association with Obama. Wright's notoriety and fame were incidental to, and derived from, Obama's fame. Wright rode Obama's wave, not by choice, but by virtue of his relationship to a potential presidential candidate and, later, the potential first black president of the United States. Wright came to light for most Americans on Obama's terms, or at least on terms that established his relationship to Obama as the main reason we should pay attention to him. Thus Wright's prophetic duties became subsidiary to his service to Obama as a pastoral figure and counselor. In both of his books Obama praised Wright, who, in no small measure, helped to establish Obama's religious authenticity.

While Wright heated up the media circuit, many Americans reviled him for seeking to clarify his views and to defend himself against gross misrepresentation. It was bad enough that the mainstream had little interest in his views before they were savagely ripped out of context; it was worse still that it had little interest in hearing the complicated reasons for his views aired without the backbeat of shrill denunciations. Wright sought to take his voice back from the media hacks who'd stolen it; he sought as well to get his name back from a press that had soiled his reputation through its ignorance and its small-minded scripts of what was appropriate for black preachers to speak about. King suffered the same fate in the press when he dared move from speaking about civil rights to foreign affairs. He was in print stripped of his Ph.D.—suddenly "Dr." was dropped from his name—dismissed as an incompetent radical, and relegated to the loony bin of marginal figures who had lost their bearings and relevance.[51]

If the press did Wright wrong, it may also look as if Obama kicked dirt in his face when he was down, but that would be a poor reading of the situation Obama faced. Wright said at the National Press Club that Obama would do what politicians do—though in his case, that is not altogether true, at least not in the instance of a politician defending a friend who had fallen on hard times. Obama's defense of Wright in his speech was perhaps not the best a politician might offer, but it was far better than what most seem willing to give. One need only think of how quickly Bill Clinton abandoned his assistant attorney general nominee Lani Guinier the moment her rather reasonable ideas got distorted by the right wing and pegged as controversial in the press. If throwing friends under the bus were an Olympic sport, Clinton might win the gold medal for a triathlon of black dissing that also included Jesse Jackson and Joycelyn Elders.

Obama insisted that he could no more disown Wright than he could disown the black church. On the one hand, it was a noble gesture. But as was true with Obama in other areas, what he gave with one hand he partly took back with the other. Obama painted the black church as containing the good and the bad of the black community—though he refused to draw a parallel with the white world. The failure to draw such a parallel turns what might be a true observation about both groups into an indictment of just one of them. Wright became an emblem of the promise and peril of black religion; before Obama's presidential run, no such worries seemed to trouble him. Wright's sermonizing had not caused Obama to exit the church.

Wright's words, and those of another minister, caught up with the presidential candidate and pushed him out of the congregation that had been his church home for twenty years. Obama left Trinity after Father Michael Pfleger, the charismatic white pastor of St. Sabina Roman Catholic Church on the South Side of Chicago, mocked Senator Clinton in a guest sermon at Trinity in late May 2008. Pfleger dabbed his eyes with a handkerchief in feigned grief and suggested that when Clinton cried during the campaign in New Hampshire, it was because white entitlement and her status as a former first lady had caused her to feel that she deserved to be president. "And then, out of nowhere, came 'Hey, I'm Barack Obama,'" Pfleger said during his sermon. "And [Clinton] said, 'Oh damn, where did you come from? I'm white! I'm entitled! There's a black man stealing my show!'" Obama condemned Pfleger's "divisive, backward-looking rhetoric," and although the priest apologized for his remarks, Obama resigned from Trinity several days later.[52]

Eyes on the Prize

As Obama made his bid for the big prize, he increasingly began to calculate the effect of black religion and his pastor by measuring their effect on him. Few could fault Obama for giving practical consideration to what church he attended and how what might be said there would be taken as a reflection of his political and racial views. Still, it is troubling that such a utilitarian view of his religious affairs could cause black folk to measure the health or direction of black life through its possible effect on Obama, but not take measure of Obama's effect on the health and direction of black life. Blacks must remain skeptical about elevating any single journey over the pilgrimage of the entire community, a possibility to which Wright's bloody sacrifice in the media might sadly point.

Obama insisted in his speech that black folk must find their path to a more perfect union by refusing to become victims and by tying their plight to that of white Americans. If more white Americans were willing, in Susan Taylor's apt phrase, to "link arms and aims" with black folk, there is little doubt that such a gesture would be proudly returned.[53] Even as white folk hid behind brutally segregated barriers, most black folk shared their souls and institutions with all comers as long as they shared the black desire to be free and fully human. Obama had a valid point in challenging black folk to reach across race to forge ties with whites. It is true that blacks have at times shunned the social solidarity that might make their situations in life better. There is greater validity still to issuing such a challenge to reluctant whites who have collectively had far greater privilege and a far longer history of directly or subtly holding blacks at bay.

The Bible that Obama reads, too, demands greater responsibility from groups that have had more privilege: "From everyone to whom much has been given, much will be required; and from one to whom much has been entrusted, even more will be demanded."[54] That holds true for privileged blacks in relation to their lesser kin, as well as for those whites who have been able to trade on their skin for security or unearned advantage. Obama's optimism about interracial cooperation may have kept him from seeing how a number of whites feel they have nothing in common with blacks and want nothing to do with them. Those whites who feel that black folk have already got the advantage through affirmative action surely will not see the need to forge alliances. And those whites who took to the streets in opposition to Obama and his policies surely will not feel inclined to join with blacks who have nothing like Obama's grace or standing—or his tolerance for white indifference or disdain. If Obama could split the difference between his optimism and his steely-eyed assessment of the responsibilities of people of color, as well as whites, in forging a healthy social compact, he might offer the nation a workable blueprint for racial progress. But his reluctance to broach the subject of race except when forced to do so makes that goal harder to achieve. His considerable rhetorical gifts spoil from neglect, and the nation suffers from his failure to mount the bully pulpit and offer wisdom and leadership.

A great deal of Obama's racial reticence may have to do with the pressures he faces in the mainstream. While he has become the ultimate symbol of American life, Obama continues to face enormous skepticism about his American identity and his patriotism. Such skepticism is not unique; it reflects persistent questions about black loyalty to the nation as black critics have historically called America to account for its misdeeds at home and abroad. How Obama has handled those pressures, and how black people, and the mainstream, respond to Obama's actions, will certainly shape debate about American citizenship, patriotism, and racial identity for years to come.

4

RE-FOUNDING FATHER

Patriotism, Citizenship, and Obama's America

IT WAS A BREATHTAKING VIEW OF MONTECITO, CALIFORNIA, A coastal enclave south of Santa Barbara dubbed "the American Riviera." Oprah Winfrey had gathered fifteen hundred folks at her palatial estate in the fall of 2007 at a fund-raiser for her "favorite guy," Barack Obama, the man she early on favored to be president of the United States of America. The invitation hailed it as "the most exciting Barack Obama event of the year anywhere." It was undoubtedly Obama's most star-studded fund-raiser to date, and Hollywood's elite hobnobbed and kibitzed: Sidney Poitier and Forest Whitaker rubbed shoulders with Cindy Crawford and Linda Evans. Obama favorite Stevie Wonder performed as Winfrey raked in \$3 million for the charismatic politico who was her Chicago neighbor.[1]

After the fund-raiser Chris Rock cornered Obama and me at dinner.

"My dad used to say, 'You can't beat white people at anything,'" Rock said as Obama and I listened intently, intrigued by his proposition. "Never. But you can knock 'em out."[2]

Obama and I chuckled as Rock gave an impromptu performance for a small but appreciative audience of two.

"Like if you have six and the white guy has five, he wins. If you're black, you can't let it go to the judges' decision 'cause you're gonna lose. No matter how bad you beat this man up."

By then Obama and I were nodding our heads in agreement. We knew the odds were often stacked against blacks in the competition to get a decent shot at a job or a seat in school. We were evenly staggered in age; Rock is four years younger than Obama, who is three years my junior. We had all likely heard members of the earlier generations say that black folk have to be twice as good as whites just to get in the arena to compete. This was Rock's version of that story, except it carried a literal punch line.

"Larry Holmes and Gerry Cooney are the perfect example of life," Rock stated. "Larry Holmes beats the shit out of this guy—for eleven rounds. He knocks him out in the eleventh round. They had to stop the fight. The man is bloody. He's been beaten the whole fight. They go to the judges' scorecards. Larry Holmes is losing the fight. If he didn't knock him out, he would have lost the title. That is essentially the black experience."

Rock's message to Obama was clear: he could not just beat Hillary Clinton, the Democratic Party's presumptive nominee, or, if he got that far, the Republican Party's candidate; he really had to clobber them with a campaign sledgehammer. As Obama soon discovered, even when he bested Hillary for the Democratic presidential nomination and soundly defeated John McCain in the general election, he still faced enormous skepticism about his patriotism and his citizenship after taking up residence in the Oval Office. Before he became his party's standard-bearer, while grappling with the Jeremiah Wright controversy, Obama had to battle the fear of some that he might secretly be an angry black man—that he might not love America after all. In this Obama may have come as close as ever to the experience of ordinary blacks who fight similar suspicions.

Critical Black Patriotism

Obama's election certainly signified new American possibilities: a new day of American hope dawned, a new meaning of American identity was born, and a new beam of governance glowed in the dark of American empire. This nation can never go back to what it meant before Obama colored the face of America for the world. But Obama's historic rise reveals flaws in the American character too. White America has shown little interest in the complicated ties of black folk to a land that has not always shown them love. Black Americans have weathered the downpour of spite from blind loyalists for their willingness to see the beauty and warts of the nation all at once. Many blacks criticize America with a patriot's desire to see the country embrace its ideals. But even a tiny blush of black criticism turns many whites red-hot with resentment and often leads to the charge that blacks do not love America. Such alleged disloyalty links black critics of the nation to the enemies of America around the globe. The same irrational fear of disloyalty prompted the nation to send Japanese Americans to concentration camps during World War II and has led to unjustifiable assaults on the civil liberties of Muslim Americans during the "war on terror" today. But black folk have loved the country and remained loyal to it even when they could not enjoy the democracy for which they were willing to die.

Claims of black disloyalty ring hollow when we consider that American citizenship rests on forced as well as free black labor, and the denial of black opportunity, in every realm. Leaders of the nation's schools and religious institutions doubted the intelligence and humanity of black folk; the Supreme Court insisted on their legal and social inferiority. The myth of American individualism flowered despite clear evidence that Americans depended heavily as a group on black sweat and also ingenuity; the minds and spirits of black people deserve as much credit as their muscle in fashioning national culture. The same cultural soil that mythic individualism sprouted from also yielded the notion that black people were irresponsible and immoral and not to be trusted with the privileges of American citizenship. The key to American citizenship is locked inside a simple formula: the more black folk were denied the rights and protections that white citizens took for granted, the more natural it seemed to those white citizens that blacks should not enjoy them. American citizenship expanded as black privilege receded. The belief that black people were undeserving of a fair share in American life cemented the civic life of white citizens and made them fully American. Undeserving and untrustworthy blacks were seen as America's greatest domestic threat; they had to be kept in line and out of democracy's limelight.

To be sure, other philosophies of governance distrust the democratic intelligence of the average citizen. But collective white guilt stirs exaggerated suspicion and fear of black citizens. In Obama's case, it flared in a crude argument before his election: Obama cannot be trusted as the ultimate symbol of American identity and power because he might do to white folk all the bad things they had done to black folk for centuries. Such fear and suspicion rested on a bad reading of black history: black people have been the least likely among those crushed by American society to seek revenge. There have been too many blacks who have shown that they revere white life far more than they revere their own culture. A significant minority of blacks have even sabotaged their race to satisfy their love of whiteness; they have betrayed black communities and causes because they feared not being loved by whites. More than a few blacks have been so taken with the graceless deceit of white supremacy that they sold out their own to preserve ties with white culture. These are the blacks during slavery, for instance, who told whites how and when slaves intended to rebel. There is simply little reason to fear that substantial numbers of black folk will ever be unfaithful to the country or even to white America.

None of this has kept blacks from trying to prove their patriotic mettle. Mainstream America has had little use or feel for the patriotism that a lot of black people practice. Black love of country is often far more robust and complicated than the lapel-pin nationalism of many white citizens.[3] Barack Obama hinted at this when he declared in Montana during the 2008 campaign: "I love this country not because it's perfect, but because we've always been able to move it closer to perfection. Because through revolution and slavery . . . generations of Americans have shown their love of country by struggling and sacrificing and risking their lives to bring us that much closer to our founding promise."[4] That is a far cry from the "my country right or wrong" credo that confuses blind boosterism with authentic loyalty. At their best, black folk battle the worst form of patriotic correctness. In its place they offer critical patriotism, an exacting devotion that carries on a lover's quarrel with America while shedding blood in its defense.

Perhaps it is easy to see why the words of black critics and leaders, taken out of context, can be read as cynical renunciations of love of nation. Abolitionist and runaway slave Frederick Douglass gave a famous oration on the meaning of Independence Day, asking, "What to the American slave is your Fourth of July? I answer a day that reveals to him, more than all other days of the year, the gross injustice and cruelty to which he is the constant victim."[5] Instead of joining the chorus of black voices swelling with nostalgia to return to their African roots, Douglass stayed put. Poet Langston Hughes grieved in verse, "There's never been equality for me, / Nor freedom in this 'homeland of the free.'"[6] But his lament is couched in a poem whose title, like its author, yearns for acceptance: "Let America Be America Again." Even Martin Luther King Jr. was called a communist and branded a traitor to his country because he opposed the war in Vietnam. When King announced his opposition in a famous Riverside Church address in 1967, journalist Kenneth Crawford attacked King for his "demagoguery," while black writer Carl Rowan bitterly concluded that King's speech had created "the impression that the Negro is disloyal."[7] Black dissent over war has historically brought charges of disloyalty, despite the eagerness of blacks to defend a democracy on foreign soil that they could not enjoy back home.

As the nation flexed its muscles as a global empire, it created an even more complicated situation for black citizens: as they were being eyed suspiciously by white citizens, America's growing global presence inspired black folk to become even more empathetic toward international struggles for human rights. The sense that they were citizens of the world often gave them the courage to fight for their rights at home. It also gave blacks moral leverage to highlight the hypocrisy of America's playing moral cop for the world while denying basic human rights to its black citizens. Blacks could now portray racism as incompatible with America's civic ideals; treating black folk right strengthened democratic culture. Black criticism of racial injustice is a valiant patriotic gesture.

When black activists criticized America for unjustly mistreating black folk, just as it mistreated other nations, they gained global allies but alienated many fellow Americans, who deemed such criticism treasonous. Black critics argued that if America wanted to preserve its democratic reputation, it had to be international but not imperial. Blacks made their native land nervous as they forged ties with other anti-imperial and anticolonial critics of American foreign policy around the world. Such efforts exposed imperialism and colonialism as racism in global drag. When blacks criticized American empire, it tore at the scab of black loyalty — and put black folk under a hyper-patriotic microscope like the one wielded by Joseph McCarthy against Paul Robeson, or the ungainly apparatus applied to Martin Luther King Jr. by J. Edgar Hoover.[8] The red-baiting of selected Americans mirrored the long-standing suspicion of black people as unreliable citizens. America's mistreatment of black folk surely made them more sympathetic to various victims of American foreign policy. Black critics were not being disloyal to America by insisting that the nation live up to its ideals at home and abroad; instead, they were exhorting the country, as Martin Luther King did in his last speech, to "be true to what you said on paper."[9] Black critics love America enough to tell it the truth and to encourage the nation to steadily enliven its democratic principles. It is not disdain but the love of America that inspires soulful criticism; as James Baldwin wrote, "I love America more than any other country in the world, and, exactly for this reason, I insist on the right to criticize her perpetually."[10]

Even the angry comments of Jeremiah Wright have to be read as the bitter complaint of a spurned lover.[11] Like millions of other blacks, Wright was willing to serve the country while suffering rejection. He surrendered his student deferment in 1963 and voluntarily joined the marines, and after a two-year stint volunteered to become a navy corpsman. He excelled at the Corpsman School, becoming valedictorian, and later a cardiopulmonary technician, and eventually became a member of the commander in chief's medical team and cared for Lyndon B. Johnson after his 1966 surgery, for which Wright earned three White House letters of commendation. Dick Cheney, born the same year as Wright, received five draft deferments, four while an undergraduate and graduate student, and one as a prospective father. Both Bill Clinton and George W. Bush used their student deferments to remain in college until 1968 and exploited family connections to avoid active duty.[12] Judged by the standard of military duty that so many claim is undeniable evidence of patriotism, Wright, much more than Cheney, Clinton, or Bush, embodies Obama's ideal of "Americans [who] have shown their love of country by struggling and sacrificing and risking their lives to bring us that much closer to our founding promise." Wright's critics have confused nationalism with patriotism.[13] Nationalism is the uncritical support of one's country regardless of its moral or political bearing. Patriotism is the critical affirmation of one's country in light of its best values, including the attempt to correct it when it is in error. In that light, Wright's words are the tough love of a war-tested patriot speaking his mind; the ability to do so is one of the great virtues of our democracy. The most patriotic gesture his nation can make is to extend to Wright and other critics the same right for which he was willing to die.

True to His Native Land

The shrinking gulf between American ideals and practices, what Obama praised as the nation's ability to move closer to perfection, led Michelle Obama to declare in February 2008 that for "the first time in my adult life, I am really proud of my country."[14] Ironically, many critics bashed the future first lady and said she was ungrateful for the good things America had done for her and her husband, especially the opportunity to thrive as a member of the black middle class and even to run for president. Black folk are nearly always charged with ingratitude and disloyalty whenever they acknowledge the tattered history of race as they offer critical love for the nation—a trait Hubert Humphrey deemed vital to America since "critical lovers . . . express their faith in the country by working to improve it."[15] The charge of ingratitude is leveled even when, as in Michelle Obama's case, blacks point to their abominable racial history to draw attention to how much progress has been made. Some Americans think that the admission that we have made errors in the past is worse than the denial of patriotism's benefits to all citizens.

During the 2008 campaign Michelle Obama was seen as a divisive force who could severely damage her husband's chances to become president. Michelle Obama had in fact simply confirmed her husband's hunch: America could correct itself and get things right once it had the courage to confront its past. She read the willingness of citizens to put her husband in the White House as a sign that such a day had arrived. Yet millions of Americans, spurred on by right-wing interests, concluded that Michelle Obama was ungrateful for her opportunities and oblivious to the progress that has been made.

Her comments revived the suspicion about black folk as disloyal citizens and untrustworthy members of our democracy. The masses of black folk, though, agreed with Michelle Obama's sentiment. She gave voice to the agony of being patriotically bipolar, of feeling, as W. E. B. Du Bois phrased it, one's black and American identities in constant tension. Michelle Obama seized the moment to speak for the black folk whose voices had only partially been heard in her husband's campaign. Barack Obama expressed the desire of blacks to be viewed as legitimate citizens, but Michelle Obama honored a political ambivalence—premised on what cultural critic Salamishah Tillet terms "civic estrangement"—that is the bedrock of black patriotism.[16]

Michelle Obama was virtually silenced after "Pride-gate" and recast by handlers as a benign presence—a woman who, despite rumors, did not secretly hate whites. If Michelle got drubbed for her comments, Barack got berated even more. Obama's patriotism was challenged early on in his first presidential campaign. When he was asked by a reporter in Iowa in October 2007 about his absent flag lapel pin, which had become standard for politicians to wear after 9/11, Obama offered unusual candor in defense of his alternative sartorial patriotism. "You know, the truth is that right after 9/11, I had a pin," Obama said. "Shortly after 9/11 . . . that became a substitute for I think true patriotism, which is speaking out on issues that are of importance to our national security, [and] I decided I won't wear that pin on my chest. Instead, I'm going to try to tell the American people what I believe will make this country great, and hopefully that will be a testimony to my patriotism." After conservative talk show hosts like Laura Ingraham and Sean Hannity pounced on him, Obama claimed: "I'm less concerned with what you're wearing on your lapel than what's in your heart. You show your patriotism by how you treat your fellow Americans, especially those who serve. And you show your patriotism by being true to your values and ideals. And that's what we have to lead with, our values and ideals."[17]

The firestorm of controversy eventually forced Obama to adopt the lapel pin as patriotic asbestos to shield himself from aggressive attacks on his love of country.

Obama's name conjured xenophobic passions during the 2008 campaign. Some opponents believed that his middle name, Hussein, suggested that Obama was an alien force out to harm the nation. Obama was blanketed by mysterious e-mails and Internet rumors that he was secretly a Muslim. Only in a climate of ignorance and paranoia about Islam in the wake of 9/11 would such charges threaten to disqualify a candidate for the Oval Office.

Bill Cunningham, a Cincinnati talk show host and John McCain supporter, played on Obama's middle name at a February 2008 campaign event by repeatedly emphasizing "Hussein" and underscoring Obama's foreignness and inherent unfitness for office. "At some point in the near future the media . . . is going to peel the bark off Barack Hussein Obama," Cunningham derisively declared. This was not a racially charged comment on its surface; the racial undertones lurked in the historic reference to Bush campaign manager Lee Atwater's infamous 1988 threat against Democratic presidential nominee Michael Dukakis: "I'll strip the bark off that little bastard and make Willie Horton his running mate." Horton is the black convicted felon who got released through a Massachusetts weekend furlough program while serving a life sentence for murder without the possibility of parole. Horton failed to return from his furlough, and nearly a year later he committed assault, armed robbery, and rape. The 1988 presidential campaign became roiled in racial drama when political ads associated the furlough program with Dukakis because Horton had disappeared during his term in office as governor. Dukakis did not start the program but strongly supported it and vetoed a bill that denied the extension of such privileges to first-degree murderers.[18] Cunningham thus scored two bigotries for the price of one when he also signified on Obama's middle name through dramatic overemphasis. To his credit, Obama's Republican opponent, McCain, immediately repudiated Cunningham's comments.[19]

Next, a widely distributed photo was alleged to portray Obama refusing to place his hand over his heart as he recited the Pledge of Allegiance. It was later revealed that the picture was taken at Senator Tom Harkin's traditional steak fry in Iowa in September 2007, where, like many Americans at sporting events, Obama kept his hands by his side during the singing of the national anthem. Less than a year later, on July 24, 2008, a teeming crowd of 200,000 packed the area around the Victory Column in Berlin's Tiergarten Park to witness candidate Obama flash his foreign policy chops and demonstrate that he had greater worldwide appeal than the sitting U.S. president. Obama was nevertheless taken to task for saying that he spoke "not as a candidate for president, but as a citizen—a proud citizen of the United States, and a fellow citizen of the world."[20] The statement indicated Obama's debt to a vibrant black internationalism that was a far cry from American imperialism. But with his own patriotism in question, many deemed it unwise for Obama to embrace his world citizenship.

Obama patiently and relentlessly fought rumors of Muslim identity and a lack of love for America. "Nobody thinks that [then-president George W.] Bush and McCain have a real answer to the challenges we face," Obama said during a campaign stop in Missouri. "So what they're going to try to do is make you scared of me. You know, 'he's not patriotic enough, he's got a funny name,' you know, 'he doesn't look like all those other presidents on the dollar bills.'"[21] The McCain campaign accused Obama of playing the race card as Obama's campaign unpersuasively insisted that Obama had merely been referring to his being a political newcomer. In truth Obama was fighting innuendo with innuendo and meeting political chicanery and racial signification on its own slippery terrain. Obama fought these forces most strongly less than a month after securing the Democratic presidential nomination in June 2008 when he spoke at the

Truman Center in Harry Truman's hometown of Independence, Missouri:

> Throughout my life, I have always taken my deep and abiding love for this country as a given. It was how I was raised; it is what propelled me into public service; it is why I am running for president. And yet, at certain times over the last sixteen months, I have found, for the first time, my patriotism challenged—at times as a result of my own carelessness, more often as a result of the desire by some to score political points and raise fears and doubts about who I am and what I stand for. So let me say this at the outset of my remarks. I will never question the patriotism of others in this campaign. And I will not stand idly by when I hear others question mine.[22]

That last line showed Obama's determination not to repeat the mistakes of the John Kerry presidential campaign in 2004 and remain silent or defenseless in the face of being "swift boated" through baseless attacks on his patriotism and character. It was Kerry who had helped Obama become a national political phenomenon by offering him the keynote speech at the 2004 Democratic National Convention in Boston. Obama learned from Kerry's experience while ratifying Kerry's belief in him as the party's next superstar.

Obama's foes tried to demonize him as a threat to the American way of life. He was bitterly denounced, sometimes serially, sometimes simultaneously, as a communist, socialist, Marxist, terrorist, traitor, and an un-American subversive. Obama frustrated his critics by becoming even more American. He spoke of hope; he chastised bitter partisanship; he said his biracial roots gave him an understanding of all races; and he claimed that his meteoric rise was possible only in America. Even as reports of foiled assassination plots surfaced, Obama continued to walk steadily toward the Oval Office with a steely resolve. In those moments he seemed to combine the swagger of John Wayne and the fearlessness of Martin Luther King.

Made in America

Few candidates of any race have run as effective a presidential campaign as Obama did in 2008. He was an extremely disciplined candidate whose focus got sharper as the campaign wore on. There were few signs of internal dissension in the ranks. Little of the inevitable jousting and wrangling that goes on in any camp made it into the press. If Hillary Clinton had had half the zip and drive of Obama's political machinery, she might well have become the nation's forty-fourth—and first female—president. Clinton began the race as the Democratic favorite, but Obama's relentless grassroots organizing, Internet campaigning, and aggressive fund-raising eventually did her in. He had taken Chris Rock's words about knocking out his opponent to heart. When Obama became the Democratic Party nominee, he stayed on message and outdid Republican Party opponent John McCain in debates, fund-raising, organizational focus, and mastery of the details. His stirring triumph over McCain suggests that the nation ignored the unfair characterizations of Obama by his opponents and saw in him the best hope for the political and moral restoration of America.

Obama's belief in America was, and remains, undeniably contagious. He proclaims our nation a land of hope awash in moral excellence and political reinvention, a place where people may dream out loud about impossible ventures and turn them into fact. He is fueled by his own improbable transformation from insecure kid to indomitable political force. Obama greatly admires Abraham Lincoln,[23] and he identifies with Lincoln's will to remake himself, and his surroundings, through imagination and hard work, summed up in a couple of simple dictums: one does not have to be trapped by a past that could be reshaped in the mirror of pitiless self-reflection; and one can take destiny into one's own hands—for Lincoln, by splitting rails, and for Obama, by toiling in communities—and then transform one's labor into constructive policies through law and politics. Obama's restless self-reinventions from community organizer to state senator, from law professor to president of the United States, make him the very image of American social possibility.

At first blush, all that Obama was not—a rich white man, the son of social royalty, a well-established Washington insider—made him a counterintuitive choice to take the helm of a political vessel that by 2008 was threatening to capsize under the weight of economic hardship, financial crisis, and the rising tide of racial chaos. Yet on closer inspection, it is precisely his mix of gifts that made him the kind of man who might right the ship of American political destiny: a remarkably sober and cool demeanor, a keen intelligence and political maturity far beyond his years, an eloquence drawn from fighting injustice, and a sensitivity to all sides of a debate born of his biracial heritage. Barack Obama's journey as a black man has not precluded his claim to national identity despite the concerted efforts of conservative opponents. Instead, it has helped to certify his standing as quintessentially American.

More than half a century ago, Langston Hughes captured the debilitating divide in the destinies of white and black children in his poem "Children's Rhymes": "I know I can't / Be President." Forty-six years after Hughes, rapper Tupac echoed that declaration: "And although it seems heaven sent, / We ain't ready to see a black president." Little more than a decade after Tupac's lament, we proved ready for a black president, and if the idea was not quite divinely inspired, though millions claim it was, the grief of dreams deferred was nonetheless lifted.

By any measure Obama's first election was a monumental day in our nation's history. African Americans were, and still are, rightly proud. The brutal facts of black existence—slavery, segregation, and the stunting of social and political ambition—have dashed the hopes of black progress time and again. The election of Barack Obama symbolized the resurrection of hope and the restoration of belief in a country that has often failed to treat its black citizens as kin. Obama's words and vision have built a bridge back into the American family for millions of blacks abandoned to social neglect and cultural isolation. Obama's historic win was the triumphant closing of a circle of possibility begun when former slaves boldly imagined that one of their kin would one day lead the nation that enslaved their ancestors. In 1968 the Reverend Martin Luther King Jr. met a bullet in resistance to his dream of equality; forty years later, Americans cast their ballots to make Obama president. Obama's election was a benchmark that helped to fulfill—and rescue—America's democratic reputation. The Oval Office is the ultimate symbol of national access to power. If the levers of influence are weighted with bias or unjust privilege, they swing away from the promise of democracy, which is America's greatest legacy. Americans of all stripes can be proud that the ideals of the founders, though trumped over the centuries by grievous instances of racism and sexism, have finally, in some profound measure, found them.

It is striking enough that a black man skirted the irritating limits of race to become the nation's advocate before the world. It is even more striking that in so doing, Obama has not interrupted but rather extended the democratic ideals of the Founding Fathers. When he took the oath of office to uphold the Constitution and to defend the nation from its enemies, Obama joined a precious short list of men who have shaped the nation's future and direction for good or ill. Obama's efforts have not solved the riddle of race in America. We are not a post-racial land, and we have hardly vanquished racism from our midst. But Barack Obama's election as president promised to bring us closer to the day when a man can be considered an American even when it is recognized that he is also black. Yet we have by no means cleared all the hurdles that litter the path of progress.

Father Knows Best?

When he was first elected president, it seemed that Obama had integrated the pantheon of figures who have inherited from George Washington the symbolic role of the nation's father. The group of men who birthed the nation without a woman's womb—a reversal of the divine miracle of the Christian God getting his son to earth without male intervention—is known by the fortifying term Founding Fathers. These figures have also cast either a bright light or a negative shadow over the nation's political landscape. Obama's interpretation of America's ideals and destiny promised to enliven the creeds that have shaped the nation's self-image. Obama loomed over the national landscape as a re-Founding Father of sorts, a symbolic patriarch who guards the American way of life and redefines American moral ambition through his speech and action. Barack Obama's virtues and strengths as a father are regularly celebrated. The Obama clan offers the nation an inspiring portrait of the American family at its finest. The image of a black man openly loving his remarkable wife and adorable daughters combats entrenched stereotypes of the dysfunctional black family.

Obama has not been shy about taking to the bully pulpit to champion black fatherhood. As a presidential candidate in 2008, he spoke on Father's Day to a black church where he urged black men to stop acting like "boys" in forsaking their paternal responsibilities. Obama repeated the diatribe in his widely discussed Morehouse College commencement address in 2013 as he implored his young listeners to make no excuses and to step up responsibly to the plate of manhood. These speeches go over well in quarters of black America hungry for moral reproof as a sign of tough love. White Americans and others who are grateful to him for saying what they would like to say in public without sounding heartless or racist hail Obama's words as examples of racial and moral courage.

Obama is undoubtedly cheered as black America's father in chief, but he has faced brutal barriers in garnering respect as what might be termed our *pater politicus,* a symbolic role of national political father that came naturally to his predecessors. Some presidents have been comforting patriarchs, like Franklin Delano Roosevelt, who gathered his fretful brood around their radios in war and domestic crisis for encouraging fireside chats to remind them that fear was not their master. Other presidents were abusive fathers, like Richard Nixon, whose paranoia and insecurity drove him to belittle the nation as his petulant offspring. There were smart but bigoted political pops like Woodrow Wilson, and magical and mysterious father figures like Abraham Lincoln, who worked feverishly to keep his national family together despite looming political divorce.

Lyndon Baines Johnson was a Big Daddy who commanded southern colloquialism and backroom dealing to keep one side of the family from running roughshod over others by offering the vulnerable a vote of confidence. Bill Clinton was a bright but mischievous rolling stone papa whose appetite for winning caused him to embrace triangulation and to welcome relations outside his political tribe. Ronald Reagan was a grandfatherly sage piloting us through social turbulence while insisting to the nation's darker peoples and regions that it was morning in America. John F. Kennedy was the nation's eternally youthful dad brimming with ideas for new frontiers, a father figure frozen in memory by murder.

Obama, too, has many of the better traits of the nation's former fathers. He cares for ordinary citizens by offering Obamacare the same way Franklin Roosevelt looked out for Americans with the New Deal. Like Lincoln, Obama seeks to rival his onetime campaign opponents by employing their ideas and, in some cases, the figures themselves, from Hillary Clinton to Chuck Hagel. And like Kennedy, Obama oozes charisma, or, more to the point, as one of my students put it, Kennedy had charm but Obama has swag.[24] And therein lies the rub: the way Obama walks and talks shouts blackness in spades, a blackness that taps ugly veins of racial discomfort and, at times, unguarded animus. Obama has been rejected as the nation's father by millions who refuse to recognize his paternity for no other reason than his race. Truman's fatherly responsibility was summed up in the motto that crowned his desk: "The buck stops here." The motto of bigots out to foil Obama's symbolic fatherhood seems to be: The buck must be stopped.

The rejection of Obama as political father began nearly as soon as he took office. Heated intolerance toward Obama's presidency poured from the spouts of the Tea Party. It often tasted like a bitter brew of bigotry that was barely sweetened by claims that their opposition was strictly political and had nothing to do with color. That is hard to swallow, since one of the Tea Party's biggest tiffs with Obama is over the swelling government and its reckless entitlement spending; yet none of that inflamed the Tea Party into action when George W. Bush had control of the nation's political and financial reins. With Bush in charge, Medicare got a $1 trillion windfall, federal education spending shot up by 18 percent, domestic discretionary spending rose more between 2001 and 2006 than during Bill Clinton's entire two terms, and Wall Street and the automobile industry got bailed out, long before Obama darkened the door to the Oval Office.

Obama's efforts to care for the uninsured were tagged as politically blasphemous, while placards at Tea Party rallies framed him as an African witch doctor. The not-so-subtle message was clear: Obama is not one of us; he is something dark and different. Obama's African paternal roots were often evoked to suggest his utter otherness, his fatal foreignness, and his essential and fundamental un-Americanness. Conservative critic Dinesh D'Souza committed a treacherous act of literary interpolation and racial legerdemain to discern in Obama's hunger for his African father's memory a dark and despotic desire to unmake America and dismantle the nation.[25] The birthers cut straight to the chase and argued that Obama's birth certificate was phony and that he was not born in America, a claim amplified by billionaire political huckster and 2016 Republican presidential candidate Donald Trump. The moral sickness of such claims was not quarantined to the loony right-wing margins but bled into the rhetoric of the conservative mainstream. Former New Hampshire governor John Sununu wished that Obama "would learn how to be an American,"[26] while former House Speaker Newt Gingrich labeled Obama the "most dangerous president in modern American history."[27]

The claims of the Tea Party, birthers, right-wingers, and conservative politicians, which I will discuss at greater length shortly, rested on the vicious premise that Obama was not authentically American, and led to the conclusion that he could not be a bona fide president. If Obama is not one of America's beloved sons, he surely cannot be the nation's symbolic father. The efforts to unbirth Obama were an attempt to unfather him as well. The conservative politicians and media figures who disrespected Obama with their intemperate fits of hubris and their uncontrolled anger—Charleston, South Carolina, congressman Joe Wilson's infamous "You lie" ejaculation at the president's speech on health care before a 2009 joint session of Congress; Arizona governor Jan Brewer's condescending finger-pointing at Obama's face on the tarmac at the Phoenix airport after he descended the stairs of Air Force One in 2012; and *Daily Caller* reporter Neil Munro's rude interruption of Obama as he made a statement on immigration in 2012 from the Rose Garden—embody the rebellion against Obama as the nation's legitimate political father.

The resistance to Obama as our patriarchal proxy among broad segments of the population is rife with irony. Obama has made a great deal of his search for his absent father's meaning in his life in the luminous *Dreams from My Father*. It is a painful absence that Obama has since tried to redress in his determination to be a better father to his daughters. The enduring memory of a father's abandonment has perhaps sparked Obama's sometimes harsh rebuke of black fathers in his speeches to black audiences. It is bitterly ironic that large numbers of Americans hatefully reject a man who has struggled so publicly to forge a redemptive paternity. Longing desperately to be a better example than the father who cast him aside, Obama faces perhaps the ultimate, and most unforgiving, contradiction: as a child he was denied his father's nurture and care, and now as the nation's father, he seeks to care for millions who want nothing more than for him to disappear.

If large portions of America have rejected Obama as father, they have also rejected his "family" members, former attorney general Eric Holder and former United Nations ambassador and later national security adviser Susan Rice. Beyond race, their kinship consists of serving in the same administration and enjoying access to elite education. Their bonds are tightened because all have endured questioning of their intelligence: Obama was painted as the unjust beneficiary of affirmative action when Donald Trump suggested that the president's college grades were not good enough for him to have been admitted to Harvard Law School. Texas senator John Cornyn belittled Holder in 2012 when he said of the attorney general, "I have not been impressed with his intelligence."[28] And Senator John McCain, who graduated near the bottom of his Naval Academy class and put forth Sarah Palin as a vice presidential candidate, argued that Rhodes scholar Susan Rice was "not qualified" to be secretary of state.[29]

Holder was from the start viewed suspiciously because he argued that the failure to engage one another fully outside our racial cubbyholes at work makes us a "nation of cowards." The attacks on Holder were even more dramatically stepped up as he made aggressive efforts to uphold the Voting Rights Act of 1965, especially when he assailed the legal basis of voter ID laws in states like Texas and Florida, declaring in 2012 that these laws were a "political pretext to disenfranchise American citizens of their most precious gift."[30] Rice's major sin seems to be that she is Obama's friend. In retribution for her political kinship, and to settle political scores with McCain, Rice's informal bid to be secretary of state was torpedoed when she stood proxy for Obama's and the State Department's handling of the crisis in Benghazi, where, on September 11, 2012, Islamic militants laid siege to the American diplomatic compound in the Libyan city and killed U.S. ambassador J. Christopher Stevens and U.S. Foreign Service information management officer Sean Smith. The association of Rice, Holder, and Obama showed, ironically, how brains and swag often evoke suspicion of black intelligence and competence. If Obama's success as the nation's first black president has been hailed as a high-water mark in our country's history, his tragic rejection as

America's father by large portions of the nation's citizenry is a black
mark of the worst kind.

Tea Time and Birthers of a Nation

Despite the incredible goodwill unleashed by Obama's election
and reelection, huge problems prevailed. Obama was accused during the 2008 campaign of letting down American troops, seeking
to teach kindergartners about sex, promoting infanticide, being a
pal of terrorists, and being a closet socialist. After his win, lingering
skepticism about Obama's character and American identity stalked
in the form of Tea Party rallies and the birther movement. Soon
after Obama took office, conservatives launched a series of national
protests to highlight what they saw as the ills of the Obama presidency: expanded government and wasteful spending, symbolized by
the stimulus package of bailouts for big businesses, increased taxes,
and Obama's proposed budget.

The self-anointed Tea Party took its name and spirit from the
Boston Tea Party, where colonial rebels in 1773 protested "taxation
without representation" and demanded tax cuts, decreased public
spending, and smaller government.[31] The new Tea Party platform
was borrowed not so much from the American colonists as it was
from the antigovernment playbook of Ronald Reagan. The protests
took place on "Tax Day," April 15, 2009; July 4, timed for Independence Day, to protest Obama's un-American practices; and then
September 11, 2009, to underscore the Tea Partyers' patriotic mettle
and to suggest that all was not quite cricket with Obama when it
came to defending American interests. Despite the Tea Partyers'
attempts to portray Obama as anti-American, and to suggest that
his policies bordered on communism—"Obama for President of
Cuba," one protest placard read—polls indicated broad American
support for Obama's $787 billion antirecession stimulus package,
and for a greater governmental role in jump-starting the fatally flagging economy.

Although billed as populist uprisings, some of the Tea Party protests turned out to be an example of "Astroturf" rebellions: well-
organized campaigns disguised as spontaneous grassroots gestures.
Conservative lobbyists with deep pockets helped to organize and co-
ordinate the protests, including right-wing billionaire David Koch,
who admitted that the Tea Party protest movement was hatched at
a summit sponsored by his conservative advocacy group Americans
for Prosperity, and by Dick Armey, former Republican House majority leader and board chairman of the nonprofit conservative organization FreedomWorks. Unelected conservative forces attempted
to mask their orchestrated "movement" and question the patriotic
legitimacy of a duly elected president.

Beyond their dissembling and political machinations, another
troubling element emerged: The Tea Partyers have often given the
impression that they were angry not simply at Obama's policies, or
even his political character, but at his color, too. Political resistance
merged with racial resentment in the Tea Party movement as ven-
omous old forms of bigotry grew new fangs. White nationalism re-
verberated inside the Tea Party even though a smattering of black
and Latino folk participated. While it would be unfair to tag the
movement as explicitly racist, it is precisely the implicit character of
its racial meanings that permits racial deniability while reinforcing
racial animus. Opposition to Obama was supposedly fueled not by
race but by political principle. Tea Party rallies were often raucous
and vented spleen at anyone to the left of Rush Limbaugh. Tea Party
members shouted down various Democratic members of Congress
when they held town hall meetings to discuss health care reform. Yet
it was hard to ignore the signs of racial hostility sprinkled amidst the
flag-waving and Tea Partying.

Some demonstrators at town hall meetings that Obama attended
brandished weapons: in one case, a protester in Arizona brought
an AR-15 military-style semiautomatic assault weapon. It is difficult
not to interpret that startling image in racial terms, since no American president before Obama has faced a similar situation. Many of

the protest signs suggested the racial character of the opposition to Obama. To be sure, taxes, government spending, deficits, same-sex marriage, abortion, and health care reform were also at stake, but the bile unleashed on Obama went beyond even that of bitter partisan politics. Of the placards, some read "Obama's Plan: White Slavery" in Madison, Wisconsin; "The American Taxpayers Are the Jews for Obama's Ovens" at a rally in Chicago; "Our Tax $ Given to Hamas to Kill Christians, Jews, and Americans, Thanks Mr. O" in Sacramento, California. Another read "Obama: What You Talkin' About Willis? Spend My Money."

In Chicago, a placard reading "Barack Hussein Obama the New Face of Hitler" featured a photo of Obama imposed on the body of the German dictator, with Hitler's distinctive mustache superimposed for good measure. In Tampa, Florida, a poster featured a black figure resembling President Obama slitting the throat of Uncle Sam. At several Tea Party rallies, and in e-mails and on websites, too, Obama was portrayed as a witch doctor festooned in a feather headdress with a bone through his nose. And at an anti–health care Tea Party rally led by conservative congresswoman Michele Bachmann on the steps of the Capitol in November 2009, one sign called Obama a "Traitor to the Constitution" and another called him "Sambo," while still another asked, "Ken-ya Trust Obama?" The racial overtones and racist undertones simply could not be missed.

The skepticism about Obama's birth unspooled in extravagant conspiracy theories among the "birthers." The birther movement was born in 2008 after Obama won the Democratic presidential nomination; it resurfaced during his historic inauguration as lawsuits were filed at each juncture to remove Obama from the ballot or to keep him from taking office. Birthers challenged the legitimacy of Obama's American citizenship and therefore his ability to hold the highest elected office in the land. Article 2 of the United States Constitution says that one must be a natural-born citizen to be president of the United States. Some birthers contend that Obama was born not in Hawaii but in Kenya, and that his birth certificate was forged.

Others claim that Obama is a citizen of Indonesia, or that he had dual British and American citizenship at birth, which means that he is somehow not a natural-born United States citizen.

These claims persisted even as the Obama campaign supplied a certified copy of his Certification of Live Birth in Hawaii on August 4, 1961. Birthers claim that the certificate was not sufficient proof since it was a copy, but the director of Hawaii's Department of Health confirmed that the state possesses Obama's original birth certificate. That assurance failed to quiet the birthers' attacks on Obama's American citizenship and identity. The effect of the attacks was not limited to the lunatic fringe; a 2009 Daily Kos/Research 2000 poll showed that 58 percent of Republicans either did not believe Obama was born in the United States or were not sure, compared to 7 percent of Democrats and 17 percent of independents. The White House released copies of Obama's long-form birth certificate in 2011, and his reelection campaign even sold "Made in the USA" mugs with a photo of Obama and the image of the birth certificate. The reelection campaign reasoned that there is "really no way to make the conspiracy about President Obama's birth certificate completely go away, so we might as well laugh at it — and make sure as many people as possible are in on the joke."[32]

Face of the Nation

While Obama endured attacks on his patriotic pedigree inside America, he became the welcome face of the nation to the world. In his first year in office, Obama traipsed across the global geography more than any president before him in a comparable period and worked diligently to restore America's standing in the international community. The country's image had been greatly tarnished by George W. Bush's penchant for turning foreign policy into an extended display of America's "strategic narcissism," the notion "that what the United States does is the most important aspect of every development" in foreign policy punditry and policymaking.[33] This includes declaring war in Iraq, a nation that had nothing to do with the attacks of 9/11, but which had captured American economic interests in the region tied to the oil market. Obama also had to make amends as much as possible for Bush's aggressive philosophy of preemption, which really amounted to "prevenge"—getting back at someone he feared might even think about harming our nation before they could act. Obama's tour of redemption began in Europe as the president acknowledged American arrogance as a flaw in foreign policy. Obama was hammered by conservative critics for admitting that America had not been perfect; they dubbed his visit to Europe "the Apology Tour."[34]

The previous administration's John Wayne bravado gave way to Obama's kinder, gentler approach. The conservatives failed to see that by admitting American error, Obama intended to reinforce American truth—that by being strategically vulnerable, he could reassert American strength. Obama knew that the rest of the world wanted to see America be a bit more humble and carry its huge influence with grace. Obama was the perfect pitchman for graceful dominance. For critics on the right, Obama did not have the necessary chutzpah and imperial swagger before the global community. For many progressives, such an approach proved Obama's seductive danger: that he was an affable figure who nonetheless carried the big stick of American power. For critics on the left, his very geniality foreshadowed his potentially troubling role as a smiling emissary for empire's dirtiest deeds.

As Obama made his way around Europe and sprinkled his share of mea culpas for America's past sins, he pledged goodwill and resources while commanding respect and demanding some show of responsibility. For instance, in an April 2009 speech in Strasbourg, Obama dramatically criticized the United States on foreign soil in a way no previous president had ever done. Speaking to a town hall crowd composed largely of four thousand students from France and Germany, Obama expressed his candid views. "In America, there is a failure to appreciate Europe's leading role in the world," he said.[35]

"Instead of celebrating your dynamic union and seeking to partner with you to meet common challenges, there have been times where America has shown arrogance and been dismissive, even derisive." But Obama also chided the Europeans for "an anti-Americanism" that is both casual and insidious, while taking them to task for failing to recognize "the good that America so often does in the world" and instead blaming "America for much of what is bad." No other American president in history could have delivered this criticism to this audience.

Obama noted that attitudes on "both sides of the Atlantic" threatened to "widen the divide across the Atlantic and leave us both more isolated" even as they obscured "the fundamental truth that America cannot confront the challenges of this century alone, but that Europe cannot confront them without America." The bold and self-confident gesture of acknowledging American flaws overrode for many critics Obama's demand that there be mutual respect between Europe and the United States. His "Apology Tour" was unconscionable to conservative critics who felt that it was treason for Obama to admit to America's failings on foreign soil. These same critics likely would not agree with Obama's national self-criticism even on American terrain. But their anger toward the president only underscored their deep doubts about his true love for America and his genuine patriotism.

Obama made a remarkable speech to Turkey's parliament in early April 2009 seeking to end the ongoing hostilities between Islam and the West by publicly embracing an ancient faith that had been soiled by fundamentalist extremists. Obama acknowledged "difficulties these last few years" and a binding trust that "has been strained, and I know that strain is shared in many places where the Muslim faith is practiced." But Obama was emphatic in his insistence that Bush's war on terror would no longer set the agenda for relations between Islam and the United States: "Let me say this as clearly as I can: the United States is not at war with Islam. In fact, our partnership with the Muslim world is critical in rolling back a fringe ideology that people of all faiths reject. But I also want to be clear that America's relationship with the Muslim world cannot and will not be based on opposition to al Qaeda." Obama pledged "broad engagement based upon mutual interests and mutual respect." He insisted that America would "listen carefully, bridge misunderstanding, and seek common ground. We will be respectful, even when we do not agree. And we will convey our deep appreciation for the Islamic faith, which has done so much over so many centuries to shape the world for the better—including my own country." Obama publicly embraced his Muslim heritage in a way he could not possibly have done during a bitter campaign that sought to portray him as a secret Muslim when he claimed that the "United States has been enriched by Muslim Americans. Many other Americans have Muslims in their family, or have lived in a Muslim-majority country—I know, because I am one of them."[36]

In June 2009 in Cairo, Obama acknowledged Western "colonialism that denied rights and opportunities to many Muslims, and a Cold War in which Muslim-majority countries were too often treated as proxies without regard to their own aspirations." Obama lashed out against the exploitation of these tensions by violent extremists that led to the September 11, 2001, terrorist attacks on the United States. He confessed that he had "come here to Cairo to seek a new beginning between the United States and Muslims around the world, one based on mutual interest and mutual respect, and one based upon the truth that America and Islam are not exclusive and need not be in competition" but share common "principles of justice and progress; tolerance and the dignity of all human beings."[37] Obama was widely praised for his balance in seeking to right the horribly off-kilter relationship between Islam and the West.

This was before Obama announced on May 2, 2011, that "the United States has conducted an operation that killed Osama bin Laden, the leader of al Qaeda, and a terrorist who's responsible for the murder of thousands of innocent men, women, and children."[38] Obama insisted that "we must also reaffirm that the United States is not — and never will be — at war with Islam . . . Bin Laden was not a Muslim leader; he was a mass murderer of Muslims." In 2014, while addressing the nation about his plans to destroy the so-called Islamic State terror organization, Obama underscored his two-pronged approach toward Islam, suggesting it had no truck with terror and that terrorists murder Muslims too. "Now let's make two things clear," he said. "ISIL is not 'Islamic.' No religion condones the killing of innocents. And the vast majority of ISIL's victims have been Muslim."[39]

A Native Son Scrambles Africa

When Obama headed to Ghana in July 2009, the tune got dissonant and the tone got much harsher. Gone was the elegant balance between acknowledging American fault and European responsibility. In its place came a stony defensiveness that put most of the blame for Africa's suffering — especially war, corruption, and tribalism — on its own poor management of its affairs. Obama left untouched the role of America and the West in the African plight. He may have admitted that colonialism's brutal legacy choked Muslim life elsewhere in the world, but he denied its role in Africa's present problems. "I would say that the international community has not always been as strategic as it should have been," Obama said in response to an interview question about whether Africa's woes resulted from a failure of U.S. policy or a failure of African governance. "But ultimately I'm a big believer that Africans are responsible for Africans."[40] He did not say the same thing about Europe; instead he generously offered American assistance while demanding European cooperation. Responsibility was shared. In Europe, Obama acknowledged colonialism's consequences and offered apologies to ward off skepticism about looming American empire.

In Africa the native's son seemed restless. Obama appeared unwilling to shoulder America's fair share of responsibility on the continent of his father's birth. Of course Obama is partly right: the horrible consequences of corruption and unprincipled leadership have unquestionably hurt Africa. But Western policies, practices, and perspectives have done great damage too, a point Obama seems to dismiss. "I think part of what's hampered advancement in Africa is that for many years we've made excuses about corruption or poor governance; that this was somehow the consequence of neo-colonialism, or the West has been oppressive, or racism," Obama said in his speech in Ghana. "I'm not a believer in excuses."[41] In an interview a week before departing for Africa, Obama, in one fell swoop, ignored the lingering effects of neocolonialism and absolved America and the entire West of any blame for Africa's predicaments. Claiming he was as knowledgeable as any American president has been about Africa—a status not difficult to achieve—Obama said he could cite "chapter and verse on why the colonial maps that were drawn helped to spur on conflict, and the terms of trade that were uneven emerging out of colonialism." Obama made quick work of that tortured history: "We're in 2009. The West and the United States has not been responsible for what's happened to Zimbabwe's economy over the last fifteen or twenty years. It hasn't been responsible for some of the disastrous policies that we've seen elsewhere in Africa. I think that it's very important for African leadership to take responsibility and be held accountable."[42]

Obama made the same points when he spoke to the Ghanaian parliament, playing up his African identity just as he had highlighted his Muslim heritage a month earlier in Cairo: "I have the blood of Africa within me, and my family's history reflects the tragedies and triumphs of the larger African story." But Obama quickly pointed to Africa's flaws and lectured his Ghanaian audience about the measures it would have to adopt to sustain American interest and stimulate American support—and satisfy American demands. "Development," Obama told Ghana's parliamentarians, "depends upon good governance. That is the ingredient which has been missing in far too many places, for far too long. That is the change that can unlock Africa's potential. And that is a responsibility that can only be met by Africans." Obama repeated his mantra that "Africa's future is up to Africans" before issuing a barely veiled threat: "We have a responsibility to support those who act responsibly and to isolate those who don't, and that is exactly what America will do."[43]

Obama's tough love for his African kin was little more than a rehash of problematic policies promoted by previous presidents. It is an error to insist that good governance has been the basis of investment and development by Western countries. Obama was essentially demanding that in order to receive aid, Africa comply with policies that benefit the U.S. government and American corporations. In sharp contrast to Obama, George W. Bush at least offered token gestures of support—including the Millennium Challenge Account and increased funding to combat AIDS and malaria. Obama simply appealed to his African ancestry to extract a political quid pro quo of African deference in exchange for American support.

Obama was also wrong in his belief that the flow of foreign investment was contingent on good governance. South Africa flourished even under the brutal regime of apartheid. And Angola and Nigeria received aid because they offered a bounty of oil in return.[44] African nations that are rich in oil also offer gas and minerals to Western corporations, banks, and governments—and plenty of opportunity to exploit weak or nonexistent democracy, corrupt government, unjust labor practices, and perpetual civil war.[45] Western governments depend upon the creation of internal turmoil and civil chaos—and unprincipled and bribable leaders in Nigeria, Angola, South Africa, Kenya, Cameroon, and Congo—to reap the biggest payoffs from their investments. Far more of Africa's wealth and natural resources flow from the impoverished continent into the coffers of rich countries and corporations than is returned to it through Western aid, investment, and trade. Good governance has too often translated into free market imperatives of the West: privatizing essential services such as telecommunications, water, and power as well as crucial social services like health and education, while taking subsidies away from small farmers and dispensing with import controls. All of these measures benefit Western governments and businesses.[46]

Obama's argument that colonialism and Western dominance have not continued to haunt Africa over the last couple of decades can be made only by someone who is willing to ignore the vast evidence of neocolonialism's present ills. True enough, the slave trade, more than a century of colonial rule, the rise of French and American neocolonialism, and the use of Africa as a pawn in the Cold War are in the past, but they are prologue to more recent practices: American and French governments bankrolled African dictators and despots and supplied arms for the vicious internal conflicts and wars that have long ravaged the continent. The rise of neoliberalism in the World Bank and IMF was tailored perfectly to the demands of a neocolonial world for more subtle policies that failed to promote growth while encouraging radical inequality. This extended the West's clear dominance.[47] Obama's trip to Ghana simply buttressed the belief that America seeks relations with Africa in order to exploit its natural resources, especially oil and gas; to use Africa as a strategic front and major battlefield in the global war on terror; and to offer America a leg up in the global competition with China for economic and political supremacy. The fact that a good number of the world's shipping lanes hug African shores make it that much more attractive.

The apparent aim of Africa policy under Obama is to reassert American superiority in light of increased competition from Europe, India, and China. European colonial powers have long been rivals in the "scramble for Africa";[48] both France and England have strong interests and colonial ties in West Africa. But upstart China has upped the ante: trade between Africa and China amounted to $10 billion in 2001; in 2008 it had increased to $107 billion, and by 2012 it surpassed $200 billion. Obama and American business followed suit on a much smaller scale in 2014 at the U.S.-Africa Leaders Summit, where U.S. companies pledged to invest $14 billion in Africa. But Obama could not resist sticking his thumb in Africa's eye when he said that the continent's nations should turn inward to solve their economic ills and stop making excuses based on a history of colonization and dependence. Obama said that "as powerful as history is, and you need to know that history, at some point, you have to look to the future and say, 'OK, we didn't get a good deal then, but let's make sure that we're not making excuses for not going forward.'"[49]

Obama once again denied American and Western responsibility for Africa's economic predicament while renewing America's bid for hegemony on both economic and military fronts. A troubling formula prevails: the less aid is offered, the more the United States must rely on its military to control Africa. America's aims are achieved through direct intervention or through the supply of arms to clients under the umbrella of AFRICOM, a relatively new independent military command that oversees all U.S. military activities on the continent—including arms sales, military training, military exercises, naval operations in the oil-rich Gulf of Guinea, and air raids.[50] Only Liberia has signed on so far. America must therefore tighten the noose on Africa and cherry-pick among nations most likely to accede to its wishes. That rationale explains why Obama chose to visit Ghana in 2009 and not his father's Kenyan homeland, which is ruled by corrupt leaders, although he belatedly journeyed to his father's native soil in 2015. Ghana was targeted because it features a relatively stable democracy and offers for Obama's purpose a far better picture of good governance—that is, after the nation recovered from the CIA-backed 1966 coup that overthrew Ghana's president, Kwame Nkrumah. Ghana, however, is not without corruption either. Obama chose Senegal in West Africa, South Africa in southern Africa, and Tanzania in East Africa on his 2013 trip to sub-Saharan Africa. He avoided Nigeria and Kenya a second time because of corruption and failed democracy while refusing to acknowledge Western colonialism's long shadow there. If America's and the West's present tenure, and lingering influence, in Africa cannot be defined as neocolonialism proper, it is certainly neocolonialism lite.

Through a Glass, Darkly

The same conservatives who took issue with Obama's apologies to Europe enthusiastically endorsed his criticisms of Africa. Obama's arguments reflect a perspective that is favored by right-wing critics: take responsibility for your own country and stop blaming the West and the United States for your troubles. If you want American and Western aid, stop the corruption. The title of an essay by Bret Stephens in the *Wall Street Journal* summed it up: "Obama Gets It Right on Africa." Calling Obama's speech "by far the best of his presidency," Stephens argued that since "British Prime Minister Harold Macmillan gave his 'Wind of Change' speech (also in Ghana) nearly 50 years ago, Western policy toward Africa has been a matter of throwing money at a guilty conscience (or a client of convenience), no questions asked . . . Maybe it took a president unburdened by that kind of guilt to junk the policy."[51]

Perhaps Obama has too quickly got free of the burdens of a history that ought to weigh on him more. Obama's blackness has cut both ways: he has been saddled with prejudices and stereotypes that unjustly hem him in while he has used his color to escape even reasonable racial obligations. The escape route is tracked in Obama's callous dismissal of Africa and his refusal to extend to his darker kin the consolation he offered in Europe to the lighter limbs of his distant family tree. Obama's blackness let him off the hook for comments that would get him accused of racial insensitivity if he were a white president. In Africa, Obama expressed ideas that undermine the progressive anti-imperialism of black activists like Paul Robeson and Martin Luther King. It quickly became clear that Obama's presidency not only symbolizes the long-delayed aspirations of black folk to become fully American but forces those blacks as well to stare into the eyes of American empire and see a black face smiling back at them.

Obama's presidency may sadly mute the black criticism of American domestic and foreign policy that has been valuable in helping the nation clarify its democratic aspirations. If American imperialism was wrong when it had a white face, it is wrong now that its skin is black. A displacement effect looms: the critical love of America that black activists have shown has been largely forced underground now that the supreme symbol of the nation is black. Obama cannot be expected to express the contrapuntal character of black patriotism—of loving the nation through rigorous and sometimes trenchant criticism. But the nation needs now more than ever a black eye and an honest tongue on the rituals of American governance and citizenship. The Obama presidency exacts a high toll when progressive blacks cannot offer this necessary gift to the nation. Black people must continue the noble tradition of critical love because their democratic potential is far from being fully realized. Their criticisms are important, too, because they remind Obama that he ultimately owes his presidency to courageous blacks and their allies who fought mightily to gain full citizenship for black folk. Without them Obama could not have become president of the United States.

Obama's actions in Africa—and, as we shall later see, closer to home—suggest that he may have grasped a cynical lesson: he looked genuinely American when he criticized blacks in Africa and in this country too. Many immigrant populations and other citizens already knew that; they have despised blackness all the way to a fuller and richer American experience. Toni Morrison has argued persuasively that America is built "on the backs of blacks."[52] It seems true as well that citizenship is often forged when Americans find common ground in faulting blacks, the ultimate other.

It may be the pressure to prove his patriotism that encouraged Obama to either ignore or scorn elements of black identity. Most ethnic groups had to fight the battle of loyalty to kin and tribe over love of country, slowly surrendering their ties to the motherland, whether Italy or Poland, and sinking roots in a country that eventually accepted them because it was proud to be a nation of immigrants. That happened when the immigrants came largely from eastern and western Europe; immigrants from Central and South America, and from the Caribbean and Africa, face greater barriers to acceptance. The promise of patriotism has burned far less brightly for African sons and daughters because their black identity was often opposed to and cut off from their American identities. The skin of Europeans promised them a relatively safe transition into American citizenship because they could slip easily into whiteness. African slaves who were brought to America in chains had a vastly different experience.

White ethnics could melt into the white mainstream, while black people were rigidly segregated. Black citizenship was ruptured, black identity fragmented. W. E. B. Du Bois brilliantly captured the tense duality of American citizenship and black identity when he wrote:

It is a peculiar sensation, this double consciousness, this sense of always looking at one's self through the eyes of others, of measuring one's soul by the tape of a world that looks on in amused contempt and pity. One ever feels his two-ness, — an American, a Negro; two souls, two thoughts, two unreconciled strivings; two warring ideals in one dark body, whose dogged strength alone keeps it from being torn asunder.

The history of the American Negro is the history of this strife, — this longing to attain self-conscious manhood, to merge his double self into a better and truer self. In this merging he wishes neither of the older selves to be lost. He would not Africanize America, for America has too much to teach the world and Africa. He would not bleach his Negro soul in a flood of white Americanism, for he knows that Negro blood has a message for the world. He simply wishes to make it possible for a man to be both a Negro and an American, without being cursed and spit upon by his fellows, without having the doors of Opportunity closed roughly in his face.[53]

Obama has become the most prominent symbol yet of a man wishing to be both black and American; for the better part of his presidency, he appears to have mastered the American side of the equation better than he has learned how to engage a healthy blackness in public.

5

THE SCOLD OF BLACK FOLK

The Bully Pulpit and Black Responsibility

I WAS WAITING OUTSIDE THE OVAL OFFICE TO SPEAK TO PRESIDENT Obama. I'd had a tough time getting on his schedule, since race was not a subject the Obama White House had been eager to embrace. I'd used all my influence to talk to the president, reminding his trusted adviser Valerie Jarrett that I had twice served as a presidential surrogate for Obama and had known him for nearly twenty years. After I'd politely declined an offer to speak to the president for ten minutes, I eventually negotiated a twenty-minute interview that turned into half an hour.

"My suspicion is that, more than anything, my election will have affected changes that have already taken place in society: different attitudes about race among a younger generation, greater comfort with diversity in positions of authority," Obama said in reply to my question about how his election had changed race in America. "I think my election reflected those changes, rather than created them. I'd like to think that my election over time will help consolidate and further fuel greater acceptance of people's differences—a greater appreciation of African American culture."

Obama believed that his presidency might positively affect the racial imagination of future generations. "Just having an African American president is something that young children, whether they're white or black, now take for granted, and that will have a ripple effect, although you can't measure something like that. It's interesting how many teachers come to me and say, 'After your election we were able to talk to a predominantly African American class differently about expectations and performance and achievement and what's possible' . . . Hopefully you add all that stuff together and it signals an improvement."

Obama is hardly naïve about the persistent features of racial inequality. "My election didn't eliminate the structure of poverty among African Americans," he continued. "It has not by itself closed the wealth gap, or the income gap between races, or the achievement gap in schools. All those require policies and long-term advocacy and determination. We haven't seen the kinds of changes yet that we need to that will be fundamental in bringing about a more just America."

The rash of racial crises during his presidency—from the Henry Louis Gates affair in 2009 to the murders in June 2015 by a white supremacist of nine black people in a Charleston, South Carolina, church—has led to calls for presidential leadership from the office's famous bully pulpit, and yet Obama has often been slow to command the rostrum to address race. "I've found in this position that it's not always true that an incident automatically triggers a useful dialogue," he told me. "What you have to do is be able to create a place where people are willing to look at things in new ways and the media is willing to look at things in new ways. As president that means I've got to pick and choose my spots effectively."

Obama's record of effectively picking and choosing his spots has been hit-or-miss; he has often sabotaged his own standards with rhetoric that is far from organic, or surprising, perhaps because it grows from controversies that compel him to react. It is understandable that Obama prefers being seen as the black *president* rather than the *black* president. But his refusal to address race except when he has no choice—a kind of racial procrastination—leaves him little control of the conversation. When he is boxed into a racial corner, often as a result of black social unrest sparked by claims of police brutality, Obama has been mostly uninspiring: he has warned (black) citizens to obey the law and affirmed the status quo.

Yet Obama energetically peppers his words to blacks with talk of responsibility in one public scolding after another. When Obama upbraids black folk while barely mentioning the flaws of white America, he leaves the impression that race is the concern solely of black people, and that blackness is full of pathology. Obama's reprimands of black folk also undercuts their moral standing, especially when his eager embrace of other minorities like gays and lesbians validates their push for justice. Obama is fond of saying that he is the president not of black America but of the entire nation. This reflects his faith in universal rather than targeted remedies for black suffering: blacks will thrive when America flourishes.

Obama's views on race feature three characteristics, in various combinations: *strategic inadvertence,* in which racial benefit is not the expressed intent but the consequence of policies geared to uplift all Americans, in the belief that they will also help blacks; the *heroic explicit,* whereby he carelessly attacks black moral failure and poor cultural habits; and the *noble implicit,* in which he avoids linking whites to social distress or pathology—or moral or political responsibility for black suffering—and speaks in the broadest terms possible, in grammar both tentative and tortured, about the problems we *all* confront. It's an effort that, as we've seen with his famous race speech, draws false equivalencies between black and white experiences and mistakes racial effects for their causes.

Targets of Missed Opportunity

When he talks about race, Barack Obama blends the voices of Abraham Lincoln and Bill Cosby. He tackles the subject in largely moderate tones, and only when he must, a nod to the careful calculation of his bearded forerunner. And like the legendary comic—before his tragic fall from grace—both blacks and whites praise him for calling on blacks to stop playing the victim. Obama's strategy of urging black, but not white, responsibility reflects his belief that white guilt is exhausted and that there will be few concessions to black demands. As Obama wrote in *The Audacity of Hope*, "Even the most fair-minded of whites, those who would genuinely like to see racial inequality ended and poverty relieved, tend to push back against suggestions of racial victimization—or race-specific claims based on the history of race discrimination in this country."[1] Obama has clearly frowned on using race-specific claims to establish racial justice.

It makes sense for Obama to shrewdly avoid race to keep from being negatively "blackened" by his political opponents. The advantage of such an evasion—one that Obama doesn't often extend to other blacks—is that it doesn't see race as the problem of its historical victims. It's as if he's saying, "Race is *your* problem, not mine." There is nobility to such avoidance. Yet such a strategy may lead us to conclude that race is not important, something Obama clearly does not believe. It also keeps the nation from learning as much as it can from Obama's black presidency. Obama cannot have it both ways: he cannot benefit from King's and the civil rights movement's efforts but fail to add in some way to their legacy—and that cannot be done by pretending that we can get beyond race by ignoring it, as the case of Shirley Sherrod proves.[2]

When the Obama administration hastily fired U.S. Agriculture Department official Shirley Sherrod after a right-winger's doctored tape of a speech she gave falsely portrayed her as a bigot, a couple of truths came into dramatic focus. It seems there was substance to Sherrod's claim that the White House had permitted conservative journalists and bloggers "to decide how to govern."[3] The White House later apologized to Sherrod for its assumption of guilt, and Tom Vilsack, the secretary of agriculture who dismissed Sherrod, offered her a new job, admitting the administration had been snookered, along with the NAACP and vast numbers of the American public, by conservative blogger Andrew Breitbart. Beyond causing momentary embarrassment to the administration, however, the incident also dramatized the harmful consequences of the gag order on race in the White House for most of Obama's tenure. In addition to the failure to offer presidential leadership on the nation's most persistent social plague, Obama's reticence has deprived America of a beautiful mind on the issue of race. It is impossible not to remember that he wrote one of the most poignant race memoirs in the nation's history, in which he offered sharp insight and lucid detail on the perilous journey to black identity. A lot of Obama's sparseness on race is in step with the nation's Herculean efforts to avoid the hard work of getting right what we took so long to get wrong. But a lot of it has to do with Obama's temperament, too, as he seeks to keep the racial peace, often at the expense of black interests.

To be sure, many reasons have been proffered for the president to avoid race, but none of them are convincing. One argument goes that Obama is very busy with big issues like continuing to strengthen affordable health care and the economy and does not have much room on his plate for race. But race shadowed the health care debate because large percentages of the uninsured are black and Latino; blacks, for instance, are uninsured at nearly double the rate of whites. And while nearly everyone was suffering in our postrecession economy, larger proportions of black and Latino communities were among the hardest hit. For instance, white unemployment decreased from its 9.4 percent peak in October 2009 to 4.7 percent in May 2015, even as black unemployment climbed from 15.5 percent to 16.5 percent before settling in May 2015 to just under 9.6 percent. During much of that time, Latino unemployment decreased only slightly, from 13.1 percent to 12.6 percent, and stood in

May 2015 at 6.9 percent. Obama can and should address big social issues that affect all Americans while paying attention to what ails black and Latino populations in particular. It is precisely because Obama wants the whole hand to function that he should not ignore how an injured thumb needs more help than a healthy index finger.

The economic inequalities fly in the face of the argument that Obama can best help black people by helping all Americans. Obama said in a 2010 interview: "I am passing laws that help all people, particularly those who are most vulnerable and in need. That in turn is going to help lift up the African-American community."[4] That sounds good in theory, but it has not worked in practice. The American Recovery and Reinvestment Act that Obama ushered into law in 2009 aimed to preserve and create jobs and promote economic recovery for those who had been hurt the most by the recession. But that money was gobbled up by states with the least racially diverse populations: Montana and South Dakota raked in about $1,200 per state resident, while diverse states like Texas, Arizona, New York, New Jersey, and Hawaii collected only $600 per state resident.[5] The "rising tide lifts all boats" approach does not work, because some communities lack economic vessels that are seaworthy, and others need help in boatbuilding through access to capital, job creation, and fair employment practices.

When I pressed Obama on the negative fallout of his "rising tide" philosophy, he argued: "We came in at a time of extraordinary economic crisis. The rising tide might not lift all boats, but a sinking tide was going to leave everybody high and dry, especially the poorest and the most vulnerable. So we had to stabilize the economy. And if you look at who got fired during the recession, it was disproportionately African Americans. Or Hispanics. If you look at who lost their health care during the crisis, it was disproportionately Hispanics or African Americans." The recovery money kept cops and firefighters and teachers employed, ensuring that the "folks who were getting hit the worst got the most help," said Obama.

"The next step then was for us to say, even as we're helping to get this economy growing again, how do we start laying a new foundation that will give opportunity to everybody?" Obama said to me. "Now, you look at something like health care; that's not a 'rising tide lifts all boats' piece of legislation. That's a 'let's make sure that folks who've been left out of the boat are included in the boat.' Which, by the way, was the criticism of those who were against the health care bill. They said, 'Eighty-five percent of the people have health care. Why do we spend all this time and money and energy focused on the fifteen percent who don't?' Which is probably the reason why it was always difficult politically and never got done."

Obama viewed health care reform as a universal program that overwhelmingly helped black and brown folk. Because everyone got health care—after all, in principle, coverage under the Affordable Care Act was universal—Obama told me it meant that "low-wage workers—disproportionately African American, disproportionately Hispanic, who are disproportionately uninsured—now are going to be able to come in with health insurance. We didn't tag that as a targeted program, because it wasn't. There's still more white folks out there who do not have health care than African Americans; but it certainly is helpful to people who, as Americans, aren't getting a fair shake."

Obama made the same argument about the billions of dollars allocated for Pell Grants and student loans. "Who gets Pell Grants?" he asked me. "I'll let you examine the statistics." I did: 20 percent of white students received the federally funded grants; 46 percent of black students received them; 39.4 percent of Latino students; 22 percent of Asian students; 36 percent of Native American students and Alaskan natives; and 26 percent of native Hawaiian or Pacific Islander students received the grants.[6]

"The point is that we are able then to make strides on issues that can close the achievement gap, or close the gap on insurance, without calling them targeted programs," Obama argued. "They are programs that help people who need the help the most. And we do that not only because it's good for those individuals; it's good for the economy as a whole. It's good for everybody. I've been trying to get out of this zero sum approach that says either you're helping black people or Hispanics, or you've got these broad generalized programs that ignore the particular problems," Obama told me. "What I say is, I'm going to create universal programs that make sure everybody's got a shot, but because it provides that ladder of opportunity for everybody, it's the people who have got the least opportunity who are going to benefit most from it. And that is consistent not only with concerns about the historically left out; but it's also consistent with what I think the broad base of Americans view as fair and just. That's their definition of equality of opportunity, which means that we can mobilize broad consensus and actually get some of this stuff done."

Obama believes that mainstream American ideas of equal opportunity are the catalyst, along with race-neutral goals, for improving the plight of minorities. His belief that universal programs provide a better outcome for minorities rests in part on political expediency: whites will embrace these programs only if they offer everyone the potential to succeed.

Obama also told me: "I do think that the discussion about targeted strategies versus broad-based strategies is probably the central fault line around which I may be criticized by African American leaders. And that particular argument is one that I'm happy to have. But I really am very confident I'm right on it," he said. "It's based not only on how I think I should govern as president—meaning I've got to look out for all Americans, and do things based on what will help people across the board who are vulnerable and who need help. But," he added, "it's also based on a very practical political reality that some of my African American critics don't have to worry about because they're pundits or they're preachers, so they've just got a different role to play. I've got to put together coalitions that allow me to get legislation through a House and a Senate that results in a bill on my desk that I can sign into law." Looking me straight in the eye, Obama rebuked the idealism of two of his most vocal black critics: "I have to appropriate dollars for any program which has to go through ways and means committees, or appropriations committees, that are not dominated by folks who read Cornel West or listen to Michael Eric Dyson." I chuckled at Obama's moxie, wondering if he was about to unload on me the way he had on West when the president collared him at a National Urban League gathering in 2010 and cussed him out for his vituperative criticism.

"I don't always have the luxury of speaking prophetically"—an arrow aimed at West's relentless and self-anointed prophetic criticism—"or in theory," a knock at my insistence that race theory be specific without being racially exclusive, that public policy be targeted toward the vulnerable. "I've got the job of governing and delivering to the people who desperately need help."

Universal Coverage

If neither I nor West is useful to Obama in the legislative process, another black intellectual looms large in the president's theory about how to reduce black suffering without aiming primarily to do so. Sociologist William Julius Wilson wrote the 1990 "Race-Neutral Policies and the Democratic Coalition,"[7] arguing that Democrats, in order to expand the party's base, should not emphasize race-specific policies like affirmative action, but should embrace race-neutral policies that overwhelmingly serve poor minority communities. Wilson has more recently changed his mind.

> In my previous writings I called for ... policies that would directly benefit all groups, not just people of color. My thinking was that, given American views about poverty and race, a color-blind agenda would be the most realistic way to generate the broad political support necessary to enact the required legislation. I no longer hold to this view.
>
> So now my position has changed: in framing public policy we should not shy away from an explicit discussion of the specific issues of race and poverty; on the contrary, we should highlight

them in our attempt to convince the nation that these problems should be seriously confronted and that there is an urgent need to address them. The issues of race and poverty should be framed in such a way that not only is a sense of fairness and justice to combat inequality generated, but also people are made aware that our country would be better off if these problems were seriously addressed and eradicated.[8]

Obama would be wise to follow suit. One size does not fit all; one solution cannot possibly apply to all cases. If one visits the local hospital, some folk are downing aspirin for headaches, some are injecting insulin to stave off diabetic shock, while others are taking chemotherapy for cancer. Policies, like medicines, are most effective when they are targeted to the social ill at hand.

Some have argued that Obama should not target black communities because he is every citizen's president and that such treatment might suggest favoritism. Obama has argued, "I can't pass laws that say I'm just helping black folks."[9] Such sentiments betray a false duality and a political naïveté. It is not that Obama should either help black folk or help all Americans. Blacks are Americans who constitute a subgroup, and a political constituency, within the nation, just as do, among others, gays and lesbians, Jews, and environmentalists. Obama has favorably responded to pressure from gays and lesbians to repeal the "don't ask, don't tell" policy that prohibited gays from serving openly in the military. Obama has yielded to pressure from Jewish advocates to boycott the 2009 UN Conference on Racism and to strike up new negotiations between Israelis and Palestinians in the Middle East. Environmentalists encouraged Obama to reduce our carbon footprint by cutting vehicle emissions and raising fuel efficiency standards. They also pressured him to attend the 2009 UN Climate Control Conference in Copenhagen, which he had planned to skip. Obama acted in the national interest in addressing the concerns of gays and lesbians, Jews, and environmentalists, since what affects them affects us all. He acts in the nation's interest whenever he addresses the plight of black Americans, who also happen to be Obama's, and the Democratic Party's, most loyal constituency. Obama's logic, like that of other critics, should be reversed. It is not that in helping everybody he helps black folk; it is that in helping black folk he helps America. Tackling race and solving the problems of the black and the poor makes America a stronger nation.

Obama's ideas about race neutrality and public policy are philosophically and politically flawed. If the universal approach were the most successful way to incorporate black interests, the ultimate expression of such universality—embodied in the Constitution, the Declaration of Independence, and the Bill of Rights—should have guaranteed the rights of blacks.[10] In Obama's view, helping all Americans helps black Americans. And yet that has proved an elusive ideal. In the 1964 Civil Rights Act, for instance, discrimination based on race was outlawed. Black folk got citizenship rights that were framed in the universal language that should have given those rights to blacks from the start because of the founders' original intent of freedom for "all men"—a questionable universality that excluded blacks and women. A huge paradox loomed: in order to establish and protect the legal and civil rights of black citizens—the same protection that had been granted to white citizens—such protection was framed in the language of universal application. By helping black folk, the entire nation was helped: the Civil Rights Act banned discrimination based on race—and on color, religion, sex, and national origin. Expanding black rights also secured the rights of many other groups.

The struggle for black rights in the sixties brought three salient facts about the universal into view. One, the universal was not a given since it had to be fought for. Two, the universal was not self-evident because it had to be argued for. And three, universality was not inalienable since it had to be reaffirmed time and again.[11]

Obama's race-neutral political strategy obscures the telling racial differences that existed in the sort of universalism he seeks to adapt to our era, even if there are unintended good consequences for black folk. But there is another philosophical conflict: if Obama is serious about bringing black suffering into focus, and relieving black pain by universal policies, the best strategy cannot be one that ignores the history of how the universal has failed black interests. The notion of the universal did not become politically viable for blacks until it was put to work in the war against racial oppression. The language of the universal in black struggle targeted black communities not by promising undue uplift but by framing politics in terms of racial history on the ground. In the past, that has meant acknowledging the racial obstacles that prevented the application of the universal. Obama's approach affirms the universal ideal without embracing the politics that should drive it. He is left with a bruising paradox: Obama aims for a racially uplifting policy by removing race from consideration. That approach is of little benefit to black folk. The proof resides in a brief empirical survey of how blacks stand under Obama.[12]

Are You Better Off?

At a press conference in 2014—where, it was proudly noted, the president fielded queries only from women, fearlessly targeting gender to make a point—true to form, blackness, but not race, took a backseat (all the reporters called on were white). When, finally, at the end of the press conference, African American reporter April Ryan forced a question about the state of black America in light of the nation's racial issues, Obama replied, "Like the rest of America, black America in the aggregate is better off now than it was when I came into office." That assertion does not hold up under even cursory examination.[13]

Under Obama, blacks have experienced their highest unemployment rates since Bill Clinton was in office. Obama doesn't even compare favorably to his immediate predecessor: the average black unemployment rate under George W. Bush was 10 percent, and under Obama it has been 14 percent. The Obama administration said or did nothing when black unemployment rose to 16 percent in 2011, a twenty-seven-year high. The economic picture is no rosier, as the median black household income dropped 11.1 percent—from $36,567 to $32,498—during the recession, more than double the figure for whites, whose inflation-adjusted decrease was only 5.2 percent. The ranks of the black poor have also swollen under Obama, from 25.8 percent in 2009 to 27.2 percent in 2012. Since education helps to combat some forms of economic inequality, the Obama administration's decision to change Parent PLUS loan requirements cost historically black colleges and universities more than $150 million and hampered the education of more than 28,000 of their students.

In Obama's administration the disparity in wealth between blacks and whites nearly doubled. The median net worth of the average white citizen is now twenty-two times the average black person's wealth—$110,729 to $4,995. Sadly, Obama has embraced policies, such as extending the Bush tax cuts, that have widened income inequality even more. The top 1 percent got tax breaks; had Obama let those breaks expire, he would have added $4 trillion of revenue to the national till. Furthermore, the *Wall Street Journal* reported in 2014 that only 1.7 percent of $23 billion in Small Business Administration loans went to black-owned businesses, in contrast to 8 percent under Bush, more than four times the rate under Obama.[14] After Obama made Bush's tax cuts permanent in 2013, he cut community block grants and gave a budget increase to the Federal Bureau of Prisons.[15]

Neither do public policies that reflect Obama's strategic inadvertence successfully offset persistent racial bias in the marketplace. Take the Fair Housing Act, the last piece of LBJ's monumental trifecta of civil rights legislation, which passed Congress in 1968 in homage to the murdered Martin Luther King Jr. and was broadened in 1988. The FHA makes it illegal to refuse to sell or rent "or otherwise make unavailable or deny" a property to any person because of race, sex, or other protected categories of identity. The Supreme Court in 2014 took up the question of whether the language of the FHA requires that intent to discriminate be proved, or whether victims of housing discrimination can claim discriminatory effect regardless of intent, ruling in 2015 that racial discrimination claims in housing cases should not be limited to questions of intent. But if public policymakers avoid the issue of race through strategic inadvertence, there is little reason to believe that housing discrimination of any sort can be eliminated without being acknowledged as structural and persistent.[16]

Even when one takes into account the unprecedented congressional obstruction Obama has faced—some of his proposed legislation would certainly have aided minority communities—the universal approach must be seen as a failure. With the shining exception of the Justice Department under Eric Holder, who left office in April 2015, Obama's approach limited the usefulness of his administration in combating racial inequality, which, in addition to the forces I have already mentioned, include higher rates of school punishment and expulsion for black kids; the disproportionate incarceration of black people; the targeting of black prisoners for capital punishment; the unequal access to high wage labor—from pipefitting and plumbing to construction work—because of the legacy of segregated unions; the stubborn resegregation of public schools; and persistent discrimination in many forms of employment, education, housing, and health care. These ugly realities expose the ineffectiveness of race-neutral policies.

Obama's failure to grapple forthrightly with race underscores a historical irony: while the first black president has sought to avoid the subject, nearly all of his predecessors have had to deal with "the Negro question."[17] Every president from Washington to Lincoln had to wrestle with slavery. Every president from Andrew Johnson to Cleveland had to deal with Reconstruction. Every president from McKinley to Lyndon Johnson had to address in some manner Jim Crow and legal segregation. And every president from Nixon to George W. Bush had to face racial and economic inequality. Some presidents even became great because they wrestled heroically with race: Lincoln helped to free slaves and thus freed America from its bondage to bigotry; and Lyndon Johnson clipped Jim Crow's wings with the Civil Rights Act and the Voting Rights Act and helped blacks and the nation to soar. Neither of these men was black, and yet they triumphed over the demons of race. Obama wishes to be treated like a president regardless of color; that cannot be so as long as he seeks exemption from the demand to address race like every occupant of the Oval Office before him. It is unfortunate that our nation's first black president has been for most of his two terms uncomfortable dealing with race; it is even more unfortunate that he could not, for the most part, openly embrace, in the course of his duties, the vital issues of the group whose struggle blazed his path to the White House.

The Bullying Pulpit

As we've seen, Obama has, in the worst way possible, targeted black people, not for support but instead for moral reproach. Even if one buys Obama's bleak version of racial uplift, there is trouble: It not only draws from the defective belief that white Americans simply will no longer accept racial victimization, a bad enough interpretation of our complicated racial situation. It also reinforces white innocence by failing to acknowledge or analyze how white society has caused and benefited from black suffering. Obama has concluded that since we are past the expiration date of white accountability—an idea that the president and millions of others confuse with the notion of white guilt—it makes no sense to keep demanding its return. That is a galling instance of racial fatalism. A wrong has been done, but we are unwilling to say who stands to gain or lose from its commission because doing so will provoke anger in its powerful perpetrators. That is not only poor social analysis and poor moral reasoning but, tragically, it is also poor presidential leadership. Obama's grating lectures, however, echo in contradiction. Even as he disavowed racial solidarity to garner votes and to govern, black pride in his presidency spared Obama the sort of lashing he routinely gives black America. Obama's bad chidings are not to be confused with a noble black moral art: pitiless self-examination which also takes the time to criticize a society that will not let black life breathe.

Obama began chastising black folk even before he found the bully's pulpit in the Oval Office. During his first campaign for the presidency, Senator Obama spoke on Father's Day in 2008 at the Apostolic Church of God, a black congregation on Chicago's South Side. When Obama assailed absent fathers for the suffering they brought black communities, he sought to gain a couple of political advantages for the price of one. He embraced a complicated tradition of social thought which says that black families are largely responsible for their own troubles. And he was seen in a black church not railing at racism but rebuking his own race—thus making a strong appeal to social conservatives. Obama's words may have been spoken to black folk, but they were also aimed at those whites still on the fence about who to send to the White House.

The modern belief that black families are mired in self-imposed trauma stems from Harvard social scientist and Johnson administration official Daniel Patrick Moynihan's infamous 1965 report on the black family. Moynihan argued that the black family was a "tangle of pathology" whose destruction by slavery festered in female-headed households, absent fathers, and high illegitimacy rates. Interestingly, Martin Luther King was one of the few Negro leaders who refused to condemn the future New York senator's report when it leaked out. If one thinks Obama was hard on black families—he said, "We need families to raise our children" and "fathers to recognize that responsibility doesn't just end at conception"—King's words might sound downright inflammatory.[18] "The shattering blows on the Negro family have made it fragile, deprived and often psychopathic," King said in conjuring a depressing image to convey his beliefs about the domestic suffering of black folk. "Nothing is so much needed as a secure family life for a people to pull themselves out of poverty and backwardness." But King insisted that Moynihan's report offered both "dangers and opportunities," the latter including the chance to gain support and resources for the black family. The danger was that "problems will be attributed to innate Negro weakness and used to justify neglect and rationalize oppression."[19]

The last fifty years of sociological diagnosis of the black family suggest that the dangers have won—and ignorance, too. A 2007 study by Boston College social psychologist Rebekah Levine Coley concludes that black fathers not living at home are more likely to keep in contact with their children than fathers of any other ethnic or racial group.[20] Coley offers a more complex and less stereotypical view of low-income, low-skilled absent black fathers than does Obama. She finds that stunted economic and educational opportunities, and with them the failure to live up to the expectation to provide for their families, drives poor black men into despair and away from their families. Such findings render arguments about black fathers' inherent pathology and moral lassitude highly untenable. These men need jobs, not political jabs.

Obama's tough talk on Father's Day embraced the Moynihan Report's admonition of absentee fathers. The presidential candidate noted in passing the need for more police and money for schools, for more afterschool programs and better teachers, and fewer guns flooding the community. But he laid most of the blame on black families and black fathers in blunt—and occasionally belittling—terms: "What makes you a man is not the ability to have a child. Any fool can have a child." (It's hard to imagine Obama similarly calling out the flaws of white fathers, much less calling them fools, especially with a vernacular spin.) "That doesn't make you a father. It's the courage to raise a child that makes you a father." That kind of frank talk has been heard before; Obama's speech is straight from the Jesse Jackson playbook of personal responsibility. "You're not a man because you can make a baby," Jackson preached in the seventies. "You're only a man if you can raise a baby, protect a baby and provide for a baby."[21] But like King before him, Jackson understood that one must overcome barriers that thwart initiative and personal responsibility. Obama brilliantly quoted a Chris Rock comedic routine about black men expecting praise for things they were supposed to do, including staying out of jail and adequately taking care of their children. But Rock's humor is effective because he is just as hard on whites as on blacks. That's a part of the routine Obama still has not adopted.

Of course that is a deliberate oversight. Obama's stinging rebuff of black fathers and his firm insistence on personal responsibility were calculated to win over white social conservatives who were turned off by Jeremiah Wright's tirades against persistent racism. The last thing Obama wanted to do was to go into a black church and say anything negative about whites or to highlight the diminished opportunities for struggling blacks. Such a gesture would have struck us as odd for a race-avoiding candidate. He most visibly got racial when criticizing black culture. Since Obama became president, little has changed.

Obama's harsh rebukes, especially the bead he draws on black fatherlessness, prompted criticism from political scientist and talk show host Melissa Harris-Perry. Obama visited Chicago in 2013 and gave a speech on gun violence and poverty in the wake of a rash of black murders in his adopted hometown, most notably the death of fifteen-year-old Hadiya Pendleton, killed a week after she performed at events in D.C. during the president's second inauguration. Obama turned to a familiar theme of father absence to explain in part the problems of black communities. He lamented that for a "lot of young boys and young men in particular, they don't see an example of fathers or grandfathers, uncles, who are in a position to support families and be held up in respect. And so that means that this is not just a gun issue; it's also an issue of the kinds of communities that we're building. When a child opens fire on another child, there is a hole in that child's heart that government can't fill. Only community and parents and teachers and clergy can fill that hole."[22]

Harris-Perry took issue with Obama's "Daddy issues." On social media, Harris-Perry tweeted: "Sigh . . . The Fatherhood thing is distressing to me President Obama. I know you don't mean to say that single moms cause gun violence, but . . ." After she was leveled on

Twitter, Harris-Perry took to the airwaves to clarify her position. She didn't back down as she finished the sentence she'd begun on Twitter: " . . . there are several reasons we need to be wary when policymakers evoke familial explanations for structural inequalities," she said, offering three reasons to bolster her argument. First, policymakers "tend to be blind to the pathologies of the privileged," while only the poor "are spectacles of concern for us." The Newtown, Connecticut, mass murderer had a single mom, Harris-Perry said, but she was left wealthy after her divorce, and the Columbine shooters, and the shooter of Congresswoman Gabby Giffords, too, were all reared in stable two-parent households. The correlation between fathers' absence and violence is questionable. "The recipe to stopping gun violence is much more complicated than 'Just add Dad,'" Harris-Perry insisted.[23]

Second, "to the extent that fatherlessness is the problem, there is very little that the president can, or should, do to create a solution," said Harris-Perry. Obama can't "make men marry the mothers of their children," and he doesn't "have the power to make men be responsible parents." And neither should we want him to, she added, in light of the frightening consequence of "giving the state carte blanche to muck around in the choices we make about how . . . to construct our families."

Third, if policymakers want to encourage stable families, especially in black and Latino communities, they could reform policies that fuel male absenteeism: "the war on drugs, the aggressive incarceration of young minority men, and the rules that bar them from voting, living in public housing, securing educational loans, or finding work long after they have served time for nonviolent drug offenses." Harris-Perry argued that responsible single moms "are the ones raising sons and daughters every day," yet 63 percent of the children who live in single-mother households live in poverty. "They are holding up their end of the American bargain that offers opportunity in exchange for effort," she said, and "our government owes them more than hopes for a husband."

In a Chicago church, and later, in a Chicago community, Obama lost his moral balance by stressing personal responsibility while slighting the forces that harm black families: huge unemployment, racist mortgage practices, weakened family and child care support for poor mothers, the displacement of black labor by technology, the political assault on early childhood learning programs, and—until his 2015 efforts—the over-incarceration of black people. If we rightly expect more black fathers to stick around to rear their children, we have got to give them greater opportunity to stay home.

Reprimander in Chief

Obama has targeted black moral failings in other high-profile appearances as well. In a 2011 speech to the Congressional Black Caucus, he offered a twist to his stale portrayal of deficient blackness, casting aspersions on black politicians whose job it is to point out Obama's failure to pay attention to black issues. Obama deployed the familiar approach of expressing light empathy as a means to launch a heavy attack: "And I know at times that [keeping the dream alive for our children and facing things we've never seen in our lifetimes] gets folks discouraged. I know. I listen to some of you all."[24] The audience laughed at what was at once an acknowledgment of others' complaints and an implicit presidential complaint about that black complaint, something Obama has done on occasion, but never as explicitly as he would that night.

"I understand that. And nobody feels that burden more than I do. Because I know how much we have invested in making sure that we're able to move this country forward. But you know, more than a lot of other folks in this country, we know about hard. The people in this room know about hard. And we don't give in to discouragement."

Obama then moved from empathy with the plight of black folk to an identification that cut both ways: he identified with the black masses long enough to get them to identify with *him*. He was acknowledging black struggle, but only slightly and amorphously, as an extension of past hurt, and mostly implicitly, which is usually the only way he acknowledges black troubles in the present, rarely naming the forces that *now* plague black folk—whether unjust imprisonment and police brutality or the bleed-off of black wealth in the foreclosure crisis. The "hard" that Obama referred to resonated with the audience of black politicians who had spent decades on the battlefield for black freedom. Obama was gesturing toward the storied history of black political figures like John Conyers, Charles Rangel, Maxine Waters, and especially John Lewis, stalwarts of the struggle for civil rights and black equality. He tapped black empathy before he turned black loyalty to subversive use against black interests: "Throughout our history, change has often come slowly. Progress often takes time. We take a step forward; sometimes we take two steps back. Sometimes we get two steps forward and one step back. But it's never a straight line. It's never easy. And I never promised easy. Easy has never been promised to us. But we've had faith. We have had faith. We've had that good kind of crazy that says you can't stop marching."

Obama further exploited the links between the black past and the black present, which he does, of course, when it's to his advantage, by drawing a parallel between his presidency—including the obstacles he confronts, and the slowed pace of change that results from bruising political struggle—and the long quest for freedom. It took time to get where we are, Obama was suggesting, and it will take time to get where we want to go: "Even when folks are hitting you over the head, you can't stop marching. Even when they're turning the hoses on you, you can't stop. Even when somebody fires you for speaking out, you can't stop. Even when it looks like there's no way, you find a way—you can't stop. Through the mud and the muck and the driving rain, we don't stop. Because we know the rightness of our cause—widening the circle of opportunity, standing up for everybody's opportunities, increasing each other's prosperity. We know our cause is just. It's a righteous cause."

Obama's words were punctuated several times by applause as he regaled his audience with rhetorical magic in the climax of his passionate oration. His direct appeal to black memory, and his acknowledgment of black suffering, while draping his political quest in the language of the movement, gave the president great traction with his black constituency. I wonder why, if it was this easy for him here, he could not evoke the same when it counted for black folk, and not for himself, his own career, his own political fortunes?

> So in the face of troopers and teargas, folks stood unafraid. Led somebody like John Lewis to wake up after getting beaten within an inch of his life on Sunday — he wakes up on Monday: "We're going to go march." Dr. King once said: "Before we reach the majestic shores of the Promised Land, there is a frustrating and bewildering wilderness ahead. We must still face prodigious hilltops of opposition and gigantic mountains of resistance. But with patient and firm determination we will press on." So I don't know about you, CBC, but the future rewards those who press on. With patient and firm determination, I am going to press on for jobs. I'm going to press on for equality. I'm going to press on for the sake of our children. I'm going to press on for the sake of all those families who are struggling right now. I don't have time to feel sorry for myself. I don't have time to complain. I am going to press on.

Obama explicitly embraced the civil rights legacy and identified with its ultimate symbol. The president derived further authority by quoting the only man who stands above him in black history, Martin Luther King Jr., while referring to his immortal "I Have a Dream" oration (although the quotation Obama cites is from King's Nobel Prize lecture of 1964) with mention of "prodigious hilltops" and "gigantic mountains," and then by referring to King's legendary last lap around the oratorical gymnasium with his famous "I See the Promised Land" speech. Obama's citing of one of the movement's greatest living symbols, John Lewis, fused past and present and laid claim to the influence of two icons. It is surely presumptuous to lecture most black people about the need to press on and be patient, but enough people in this audience had given speeches to know how the moment can grab hold of you and yank you by your rhetorical collar and send you flying over the rafters of normal experience way into the upper bleachers of emotional catharsis. Plus the best black traditions of moral exhortation often call on black folk to remember their past and to push forward into their future. But then the boom was lowered and reason collapsed in Obama's closing remarks: "I expect all of you to march with me and press on. Take off your bedroom slippers, put on your marching shoes. Shake it off. Stop complaining, stop grumbling, stop crying. We are going to press on. We've got work to do, CBC."

In the style of the best black orators, Obama had gifted the audience with golden eggs of black rhetoric. But then, just as quickly, he crushed what he had delivered with profound disrespect. Some of the legends he addressed were getting their skulls cracked when Obama was an infant; many of them, when he was a youth, had resolved to carry black grievances to the powers that be. It was their job to confront authority and to demand that presidents address sky-high black unemployment, but Obama was now lashing back. While scolding black politicians for doing what they should be doing—which consisted, in part, of telling him what *he* should be doing—he got angry and told them not to do their jobs. Obama suggested that these stalwarts were somehow being disloyal to a tradition that he had rhetorically mastered but whose substance he had noticeably slighted.

Obama, however, was the one, that night, who was disloyal to a tradition of critical engagement with power that he had just invoked. It is useful to remember that King criticized Lyndon Johnson even though he had been his friend and partner. Obama got angry at black leaders who simply wanted to hold him responsible in the same manner he sought to hold the black masses responsible. Obama's rhetoric of rebuke was especially painful for black leaders to bear. Maxine

Waters argued that Obama wouldn't speak to any other group, such as Hispanics, Jews, or members of the gay and lesbian community, in such a disparaging way. "I found that language a bit curious," she remarked, "because the president spoke to the Hispanic Caucus, and certainly they are pushing him on immigration . . . He's appointed [Justice Sonia] Sotomayor to the Supreme Court, [and] he has an office for excellence in Hispanic education right in the White House. They're still pushing him, and he certainly didn't tell them to stop complaining."[25]

Obama delivered a similarly dispiriting message when he gave the 2013 commencement address at Morehouse College. Obama scolded the proud black male graduates on a day meant to celebrate their singular achievement. "Too many young men in our community continue to make bad choices," the president told them. "Growing up, I made a few myself. And I have to confess, sometimes I wrote off my own failings as just another example of the world trying to keep a black man down." Obama threw down the gauntlet before these young black men: personal failure blamed on racial barriers is often a false charge made by black folk, he warned them. It was a dose of bad faith offered by a man too smart about race to talk that way, especially to a group of black males graduating from the venerable institution that produced Martin Luther King Jr. and the first black mayor of Atlanta, Maynard Jackson. Obama hammered his point home:

> One of the things you've learned over the last four years is that there's no longer any room for excuses. I understand that there's a common fraternity creed here at Morehouse: "Excuses are tools of the incompetent, used to build bridges to nowhere and monuments of nothingness."
>
> We've got no time for excuses—not because the bitter legacies of slavery and segregation have vanished entirely; they haven't. Not because racism and discrimination no longer exist; that's still out there. It's just that in today's hyper-connected, hyper-com-

petitive world, with a billion young people from China and India and Brazil entering the global workforce alongside you, nobody is going to give you anything you haven't earned. And whatever hardships you may experience because of your race, they pale in comparison to the hardships previous generations endured—and overcame.[26]

It was odd for Obama to scold young black men at an institution never known to court underachievement. To say "there's no longer any room for excuses" implies that such excuses have been regularly made. To be sure, all young folk at one time or another make excuses, regardless of the institution they attend, but that recognition challenges Obama's racially specific claim, one he seems willing to offer only with generous amounts of scorn. Obama reaches even further into the bag of black digs by drawing a bizarre connection: he says that there is no tolerance for excuses, not because slavery, segregation, racism, and discrimination have gone away, but because competition in global labor markets makes getting something you did not earn impossible. If Obama had been a student of logic, he would have failed miserably. If one were to chart this argument in a syllogism, it might go as follows: Excuses by black folk are unacceptable. Global competition for resources precludes making excuses and receiving benefits you did not earn. Therefore black folk who cite racism are making excuses for demanding what they have not earned. In Obama's fearful racial symmetry, claiming an unearned benefit on account of race is the same as making an excuse for failed performance. Beyond his twisted reasoning lies the truth that is the opposite of what Obama is saying: blacks and other minorities are often denied the goods and benefits that they would have earned were it not for racist restrictions and structural obstacles.

It is odder still for a man who garnered an honorary degree—and thus was one of the few men receiving degrees that day who had not, strictly speaking, earned it—to begrudge honor to the men of Morehouse who took home diplomas won by blood, sweat, and tears. The president's by now tired diatribe against contemporary black folk who don't measure up to earlier black people ("And whatever hardships you may experience because of your race, they pale in comparison to the hardships previous generations endured—and overcame") is a risky ploy for Obama. When he was told that neither he nor his race-averse political compatriots measured up to black leaders before them, he claimed to be part of the Joshua, not the Moses, generation, suggesting that the earlier group had its calling, and the younger cohort had one of its own. Yet Obama gives no such quarter to the generations trailing his. They may not face the obstacles of the sixties, but neither have they inherited the privileges of the Joshua generation. Obama fails to mention this crucial point: the Joshua generation may be the last black generation for a while to do better than their parents. This new generation also endures a challenge that earlier generations didn't confront: public shaming by a black president.

Obama would not, and did not, speak the same way when he addressed the Naval Academy or Notre Dame—or Barnard College, where Obama did not chide the female graduates for offering excuses because sexism in global markets might keep them from earning the same money as their male counterparts. The ready reply to criticism of Obama's tough way of speaking to black folk is that one often speaks to one's own group in ways that outsiders cannot. The caveat, however, is that one is responsible for speaking not simply *to* one's group but *for* it as well. Obama has shirked that duty by declaring he is "not the president of black America"—as if African American voters were naïve enough to believe that he represented only black interests. Obama's disclaimer cannot obscure an inarguable fact: he may not be the president of black America, but he is the president of black Americans. He owes blacks no more, but certainly no less, than he owes all citizens. Yet by endorsing the role of racial arbitrageur—one who trades in the meanings of race inside the group and criticizes it like no outsider can—Obama is also obligated to speak for the group in right measure and time, something he has rarely done. Instead, as Ta-Nehisi Coates argues, Obama routinely chews out black communities as "the scold of 'black America.'"[27]

Revisionist History and Selective Responsibility

Obama mounted the rostrum at the fiftieth anniversary celebration of the March on Washington on August 28, 2013, and let loose on black America again. I sat in the VIP section a few rows behind Al Sharpton to listen to Obama talk. Presidents Carter and Clinton had already spoken, and so had Congressman John Lewis, the only surviving speaker from that glorious day a half century ago. After acknowledging the progress and obstacles since the 1963 march, and taking pains to show that black folk's quest for freedom was identical to the struggles of most Americans—Obama usually cheers exceptionalism when it waves an American flag, not when it boasts a black face—he gave in to the lust for black reproof:

> And then, if we're honest with ourselves, we'll admit that during the course of 50 years, there were times when some of us, claiming to push for change, lost our way. The anguish of assassinations set off self-defeating riots.
>
> Legitimate grievances against police brutality tipped into excuse-making for criminal behavior. Racial politics could cut both ways as the transformative message of unity and brotherhood was drowned out by the language of recrimination. And what had once been a call for equality of opportunity, the chance for all Americans to work hard and get ahead was too often framed as a mere desire for government support, as if we had no agency in our own liberation, as if poverty was an excuse for not raising your child and the bigotry of others was reason to give up on yourself. All of that history is how progress stalled. That's how hope was diverted. It's how our country remained divided.[28]

This is a remarkably disquieting passage for several reasons. The president trafficked in half truths during the commemoration of the most visible black mass mobilization of the sixties. Obama's choleric speechifying tainted a day that should have celebrated the golden reach of King's majestic oratory across the decades to embrace the present moment in its halo effect. Although he is a gifted intellectual who understands the epic sweep of black history and the paradoxes of American politics, Obama was that day a poor public historian and social analyst.

It is perhaps boilerplate for any speech that takes note of the fits and starts of progress to argue that black folk lost their way in the perilous march to freedom. If Obama were to say the same thing to American Jews, or to the Daughters of the American Revolution, or to the marines, it would suggest that all social movements suffer from internal contradiction. Reserving such talk for black folk feels less general and more targeted, less descriptive and more punitive. Assassinations did lead to uprisings and rebellions—but so have athletic events led to drunken white men looting and rioting without commentary from the black White House. The public killing of black leaders was meant to quell social mobility and crush the desire for black progress. Assassinations sent the message that black folk should stay in their place and not share in the fruits of a democracy they helped to grow.

The assassination of Medgar Evers was meant to discourage the black will to freedom and the right to vote. The assassination of Martin Luther King was meant to murder the movement for racial justice by slaying its most visible spokesman. It was fear that he might suffer the same fate as Medgar and Martin that kept some black folk from casting a ballot for Obama during his first presidential run. Only when Michelle Obama noted on the television news magazine 60 Minutes in February 2007 that, as a black man, her husband could be shot going to the gas station did some of those fears abate and black folk flock to Obama's campaign. The assassinations of black leaders should illustrate the terror and rage they unleashed rather than figuring into a president's effort to take black America to the woodshed.

Beyond that Obama is historically off: many race riots in American history involve whites attacking blacks.[29] As for minority participation in riots, the Kerner Commission report, which grappled with a number of rebellions that dotted the urban landscape in the sixties, concluded that riots were sparked by economic injustice, racism, police brutality, geographic isolation, and the exacerbation of class inequalities. Obama's reluctance to confront law enforcement means that his brief acknowledgment of police brutality in this 2013 speech was important, though again, he quickly wipes it away with a focus on black excuse-making for criminal activity.[30] The president is suggesting a link between legitimate grievances over police brutality and black criminal behavior, a link he fashions in the wake of current racial crises involving black folk and the police. But the president might as easily reprimand the police for using black outrage over police misconduct as an excuse to harass or harm American citizens.

Obama is on shakier ground when he argues that racial politics consists primarily of the message of unity and brotherhood. The black freedom struggle pushed for social justice and radical democracy, and for a greater share in the nation's goods and resources, and its privileges, too, for which black folk had been willing to die in foreign wars, and in the struggle of America's better half against its worst in the Civil War. The push for justice often requires broad disagreement with the status quo, and hence a productive disunity that points the way to true social transformation. The unity Obama envisions cannot rest on the suppression of difference as the price of cheap togetherness or false brotherhood. True unity emerged in our nation when we came to grips with competing claims of justice and changed both the law and the array of social habits that feed democracy. Laying the burden of reconciliation at the feet of black victims of injustice is already wrongheaded; but blasting blacks for defending themselves against deadly assault is intellectually dishonest. Obama's charge of black recrimination in the face of withering attack sounds like the president evoking the angry black bogeyman as the foil of his suspect racial narrative.

Obama's next argument in his March on Washington anniversary speech is particularly dumbfounding and no less insidious. He says that the quest for equal opportunity, which he sees as the chance to work hard and get ahead, became a ruse to secure government support while denying black people agency in their own liberation. This led to black folk making poverty an excuse to neglect their children and bigotry a reason to doubt themselves. Obama has insisted that access to the American dream rides on the willingness to work hard. But this permits the president to sidestep a larger problem: the systematic exploitation during 250 years of slavery of black workers, who then became victims of peonage from Reconstruction well into the Jim Crow era.[31] Exploitation of black labor continued into the sixties, when black workers were unfairly paid lower wages than whites for doing comparable work. Martin Luther King Jr., in fact, died struggling for a living wage with striking sanitation workers in Memphis, Tennessee. Obama, in offering his interpretation of opportunity that day, had to willfully ignore the brutal history of hardworking black Americans' being denied what was justly theirs by the government that he now headed.

Obama's vision of opportunity is historically awkward as well. The president unjustly lays into black failure by focusing on equality of opportunity—a code word in the affirmative action debates about whether opportunity or outcomes should be guaranteed. Conservatives often stress the former; liberals often favor the latter. Obama thus slights the racist denial of opportunity to black folk that fueled the black freedom struggle. He reinforces stereotypes of the supposed black unwillingness to work hard and the desire to live on the dole by suggesting the exploitation of government support. To do this, Obama must choose to overlook the history and effect of black agency that has been critical to black progress.

In his famed race speech in 2008, Obama lashed out at blacks for being unwilling to work with whites to enliven their social agency.

One may disagree with black nationalists who seek to end black suffering by their own striving, but the lack of agency is not a claim one can reasonably make of them. Black agency is often the only viable means to bring about black liberation; long before the state was an ally, and was instead an enemy of black interests, black folk exercised individual and collective agency to shape their destinies and realize their social ambitions. The attack on black agency has often led to unearned and unjust white opportunity in the mainstream; the suppression of the black ability to work, to own property, and to prosper financially engendered white resentment of black agency where it did manage to arise. That resentment took flame in the white torching of Black Wall Street in Tulsa, Oklahoma, in the early twentieth century, and burned in other forms of terror to punish blacks for their success. Whites routinely rioted against blacks, a fact that Obama dares not mention.

Obama leans heavily on conservative arguments when he claims that black folk use poverty as an excuse not to raise their children properly. Obama overlooks the difficulties that poor parents confront in rearing their offspring: they often work multiple jobs and therefore cannot make every parent-teacher conference; they cannot spend as much time helping with homework because they must labor to keep the family together. So they suffer lower rates of educational attainment; and they experience depressed wages, inadequate health care, and other forms of social distress.

It is odd, too, that Obama says confronting bigotry is no reason to give up on oneself when the oppressed are often encouraged to dislike themselves in a culture that has already done the same. Black youth face enormous hurdles in holding fast to vibrant self-confidence in an environment that is continuously eroding black self-esteem and the opportunity that feeds it. The selves that young black folk shape in the mirror of a society that fears and despises them are selves that are tutored to surrender hope in themselves. Those selves must also combat the notion that blackness is all pathology and disease, while confronting a pressure to achieve that is especially cruel for growing youth who often lack the resources to effectively reach for success—something their parents and elders have not always attained. When black self-confidence flares in dramatic, even exaggerated measure, blacks, whether in the boardroom or on the gridiron, are deemed cocky. Outside of sports, where a cocksure attitude is often rewarded, black self-confidence can be deadly when police, who have long ago given up on presuming the innocence of black youth, put bullets in their backs or brains.

Obama's claim that it is black folk who have stalled progress, diverted hope, and divided our country is an astonishing endorsement of the notion that blacks are their own worst enemies, that their actions have blocked their own path to progress, that their bad behaviors are the biggest threat to black flourishing. This lets the broader society off the hook for anything it has done to hamper black progress, or the things it has failed to do to enable black thriving. By blaming blacks, Obama doesn't have to come to grips with the persistence of racial disparities that he has largely relegated to the sixties. Most telling, Obama's presidency, made possible by the progress forged in the crucible of black hope, a hope that united progressive elements for critical moments in the nation's history, repudiates the logic and tenor of his comments.

Obama is forced to exaggerate black responsibility because he must always underplay white responsibility. He once tried to assail white folk as a presidential candidate when he declared to a tony San Francisco fund-raiser that "it's not surprising" when whites in Pennsylvania facing tough economic times "get bitter" and "cling to guns or religion, or antipathy to people who aren't like them, or anti-immigrant sentiment, or anti-trade sentiment, as a way to explain their frustrations."[32] All those who believed that because Obama is half-white he could be even half-critical of whites in flashing the same tough love he routinely shows to blacks got a rude awakening. Beating up on your own works only when you are perceived as a member of the group; despite newfound pride among many whites in his biracial roots, Obama was generally seen as the black man whom he'd been phenotypically profiled as for most of his life. That sort of tough love is neither expected nor well tolerated among most whites, at least not in public. When Obama ambushed black fathers at the Chicago church, it was Jesse Jackson who had to apologize for wanting to slice crudely into him for his clumsy characterizations of these men. Obama wasn't nearly as tough on white folk from Pennsylvania; there was far more empathy in his tone and his remarks, but it was he who got beaten down in the press before being forced to apologize. He learned again that one gets loud kudos for bravery when one thumps poor black folk in the face. But when one leans into the metaphoric chests of whites, one gets pilloried as an elitist, or even as a racist.

Obama was reminded that even occasional expressions of empathy for black life risked being painted as unjust favor for his own tribe. The case of Trayvon Martin, the unarmed black teen killed in Florida by neighborhood watch volunteer George Zimmerman—in self-defense, he claimed—seared the nation's conscience.[33] It also brought front and center the rage and fear of black folk at being racially profiled, and being subject to deep cultural suspicion. When Zimmerman had yet to be arrested and charged with a crime, a fact that outraged black communities and tore at the scab of unpunished crimes against black people, Obama, taking measure of the profound suffering and dissatisfaction in black America, weighed in on it in very personal terms—but only after being pressured by black critics to speak up.

"I can only imagine what these parents are going through," Obama remarked, when asked about the Martin case at a press conference in the Rose Garden to announce his nomination of Jim Yong Kim for president of the World Bank. "And when I think about this boy, I think about my own kids. And I think every parent in America should be able to understand why it is absolutely imperative that we investigate every aspect of this, and that everybody pulls together—federal, state and local—to figure out exactly how this tragedy happened." Obama moved from the universal to the particular in expressing empathy for Martin's mother and father and all black parents. "But my main message is to the parents of Trayvon Martin. If I had a son, he'd look like Trayvon," Obama said. "I think they are right to expect that all of us as Americans are gonna take this with the seriousness it deserves and that we're gonna get to the bottom of exactly what happened."[34]

Obama was predictably lambasted by the right wing. Sports journalist and cultural critic Bernard Goldberg complained that "there was no good reason for the president to say if he had a son he would look like Trayvon Martin," charging Obama with "needlessly implying that it's dangerous being a black kid in America when white people with guns are around." Goldberg even fantasized about how Obama might revise the speech he offered the nation in the aftermath of Zimmerman's being found not guilty. Instead of criticizing white folk for clutching their purses in elevators when young black men enter, or locking their car doors when black kids get too close, Obama should have said, according to Goldberg: "I implied that their only 'crime' was being black. What I should have added is that there's a good reason for all of that. People—and not just whites—are suspicious of young black men because young black men give them plenty of reason to be suspicious."[35]

Obama would incur even greater criticism as he confronted black outrage over a rash of killings of unarmed black people by mostly white police.

DYING TO SPEAK OF RACE

Policing Black America

"THE GATES SITUATION WAS INTERESTING," PRESIDENT OBAMA said to me in the Oval Office. He was referring to the arrest of Henry Louis Gates Jr. and the furious debate about race and policing that it provoked, especially after Obama was quizzed about the arrest at the close of a 2009 press conference on health coverage legislation.

"I was responding in shorthand to a question that was posed during the press conference, and when you respond in shorthand on issues of race, it poses a great danger. The reason my speech in Philadelphia [addressing the fallout over Jeremiah Wright's "God damn America" comments that came to light in the 2008 presidential campaign] was successful was because I was able to round out the issues. But in the prism of twenty-four/seven cable news, you don't get to round out things. So you've got the Gates affair; I make a comment, and suddenly everyone was traveling all the well-worn arguments that had been developed since the sixties about police and African American males."

It would hardly be the last time Obama commented on a hostile encounter between black folk and the police, but compared to the incidents that followed, the Gates debacle was surely the least deadly. By conjuring the phrase "well-worn arguments" from the sixties, Obama was signaling once again that he preferred conciliation to confrontation in racial debates.

The president went on to say: "So folks in the African American community are thinking back to all the stories they heard from their grandpa, uncles, fathers . . . of being stopped [by the police]. Old folks who identified with the police officer are thinking of the dangers that police officers have to deal with and how the decline in order in cities and rising crime rates [forced] their families [to] move out. So it tracked these old arguments, and it wasn't going to illuminate. What it was going to do was just dig everybody in."

Obama drew a false racial equivalence between white fear and black suffering: police brutality, which has stalked black folk for a century, is hardly a relic of the sixties; and white flight is driven more by the perennial goad of revulsion to sharing social space with blacks than the fear of black crime.

Obama elaborated on his regret over not having embraced a format that would have yielded more insight about police and race in the Gates affair: "Rather than just give a quick two-line answer at a press conference, maybe I would've said, 'Let me get out the facts.' Once all the facts were out, then maybe I'd make a twenty-minute speech on it, or a half-hour discussion with some students that was televised. And that might've been a better way to do it." Obama felt obligated to continually seek "opportunities to talk about [race] where it doesn't look stilted, it doesn't look artificial, but it doesn't also just become some media feeding frenzy where there's a lot of sound and fury but it doesn't signify anything."

The Gates case touched an extremely sensitive nerve in the country, one that snakes through black communities and the largely white police forces that serve them, and at times scare and terrorize them too. Gates is one of the nation's most famous scholars, but his case offers an example in microcosm of many encounters—often lethal—between ordinary black citizens and the police, starting with the conflicting narratives of how the event unfolded. Sergeant James Crowley claims that he arrived at Gates's house and asked Gates to step outside, and Gates refused, at which point he entered the home and requested Gates's ID, which he did not initially produce, and that finally he was forced to arrest Gates when the professor followed him outside, "exhibiting loud and tumultuous behavior." Gates allegedly shouted, "Is this how you treat a black man in America?" and "You don't know who you're messing with." Gates says that he showed the officer his ID, demanded that the officer identify himself, which he did not do, and that he then followed the officer outside to get the policeman's name and badge, at which point he was arrested by the gaggle of police who had gathered.[1]

Several features of the story suggest lingering bias. A black man in a tony neighborhood simply seems out of place, even to his neighbors. Had a white professor trying to get inside his home called on his driver to help him jimmy his door open, he might not as readily have aroused suspicion. And when police arrived to check out the premises, they probably wouldn't have been nearly as quick to believe the worst about a white occupant clearly not engaged in a criminal act. Whatever one believes about what happened, Gates did not receive the benefit of the doubt, a reasonable expectation, since he posed no visible threat. Gates also seemed to be the victim of a police mentality that chafes at a challenge to implicit police authority, especially if that challenge comes from a person of color. How dare black folk believe that, regardless of their station or privilege, they have permission to speak back—or, as the arresting officer, Sergeant Crowley, saw it, to speak "black"—to state-enforced authority?

The Gates incident might have nudged Obama to renew his campaign pledge to get rid of racial profiling—or to puncture the illusion that his success represented a post-racial America. But his comments about the Gates affair at the press conference reaped a whirlwind of controversy. Obama said: "I think it's fair to say, number one, any of us would be pretty angry; number two, that the Cambridge police acted stupidly in arresting somebody when there was already proof that they were in their own home; and, number three, what I think we know separate and apart from this incident is that there's a long history in this country of African Americans and Latinos being stopped by law enforcement disproportionately. That's just a fact."[2]

The police got upset, supposedly because Obama had spoken out against law enforcement instead of helping cover their flanks, but also because, subconsciously, it is perhaps awfully tough to hear a black man, even the president, describe a white man as acting stupidly. Obama himself has been racially profiled by bigots and assorted right-wingers for PWB—Presiding While Black. And yet the political takeaway for Obama, still early in his first term, seems to have been a studied racial caution, when not violently forced out of it. Obama eventually invited Gates and Crowley for what was dubbed a "beer summit" at the White House to calm tensions.

But little the president did could quell the toxic situation that was seething between the police and black citizens in cities across the nation. As he lamented in 2015, a fatal interaction between blacks and the police "comes up, it seems like, once a week now, or once every couple of weeks."[3] Playing a game of racial catch-up—mirroring Obama's foreign policy doctrine of "leading from behind"—makes the president more reactor than leader, more racial barometer than thermostat. Jelani Cobb argues that the "man who once told us that there was no black America or white America but only the United States of America has become a President whose statements on unpunished racial injustices are a genre unto themselves."[4] That genre teems with hesitations and hiccups, as much as it contains insight and gravitas, and reveals Obama's tortuous evolution on race during his years in the Oval Office.

The Fire This Time

Obama faced one of his gravest racial tests when the fires of Ferguson, Missouri, roared after a grand jury failed to indict white police officer Darren Wilson for killing unarmed black youth Michael Brown. From the start, most blacks were convinced that the case would not be fairly considered by Ferguson's criminal justice system. There were doubts that the prosecution and defense were on different teams. The prosecutor, Robert McCulloch, looked as if he were coaching an intramural scrimmage, with the goal of keeping Officer Wilson from being tackled by indictment. The trove of documents released after the grand jury reached its decision included Officer Wilson's four-hour testimony, in which the six-foot-four-inch, 210-pound cop said that his encounter with the six-foot-four-inch, 292-pound teenager left him feeling like "a five-year-old holding on to Hulk Hogan." Wilson betrayed the extent of his feeling for the slain youth's humanity when he used the impersonal pronoun "it" in claiming that Michael Brown looked like a "demon" rushing him.[5] To many blacks, Brown's height and weight gave him a fighter's chance of surviving a battle with a cop as big as Wilson. To the police officer and many whites, Brown was the black menace writ large, the terrorizing phantom that stalks the white imagination. These clashing perceptions underscore the physics of race, in which an observer effect operates: the instrument through which one perceives race — one's culture, one's experiences, one's fears and fantasies — alters in crucial ways what it measures.

The novelist Ann Petry vividly captures this observer effect in her 1946 novel *The Street,* in which the African American protagonist Lutie Johnson remarks that racial perceptions of blacks "depended on where you sat." That is, if "you looked at them from inside the framework of a fat weekly salary, and you thought of colored people as naturally criminal, then you didn't really see what any Negro looked like," because "the Negro was never an individual" but "a threat, or an animal or a curse." After a black man is killed in a failed robbery, she notes that a reporter "saw a dead Negro who had attempted to hold up a store, and so he couldn't really see what the man lying on the sidewalk looked like." Instead he saw "the picture he already had in his mind: a huge, brawny, blustering, ignorant, criminally disposed black man."[6] Our American culture's fearful dehumanizing of black men materialized once again when Wilson saw Brown as a demonic force that had to be vanquished in a hail of bullets.

If President Obama's comments on race in the anguished aftermath of the not-guilty verdict in the George Zimmerman trial gleamed with light, his words on the rage that battered Ferguson, Missouri, were shrouded in darkness. They revealed a gifted leader whose palpable discomfort with race has made him a sometimes unreliable and distant narrator of black life. Obama's twin strategy of the heroic explicit and the noble implicit was on display as he spoke twice in the aftermath of the Ferguson grand jury's decision in November 2014. Earlier, when Obama gave his first statement on the cataclysm in Ferguson at a press conference on August 14, 2014, he'd been cautious to a fault. The president understandably did not want to fan the violence. But the pressure mounted for Obama to say something after the rage in Ferguson turned to fire. If Obama felt and looked weary at the prospect of repeating himself — "I've said this before," he reminded us — it hardly matched the moral weariness of black victims witnessing history tragically repeat itself.[7] Like a Hollywood film franchise, race in the United States, especially police violence against blacks, is haunted by sequels: the locations may change, the actors are different, but the story remains the same.

Given Obama's extraordinary talent for talking the nation through tough economic or political times, his remarks on Ferguson were extremely disappointing. Obama justified his reluctance to say too much by claiming he did not want to put his "thumb on the scales one way or the other." The president was right about the need to let the Justice Department's investigation run its course. To no one's surprise, the DOJ eventually found that it couldn't meet the high bar for bringing civil rights charges against Wilson, though it found plenty of fault in the racist practices of the Ferguson Police Department. But one cannot ignore how badly the scales of justice have been tipped against the residents of Ferguson, and how Lady Justice has had her blindfold removed and discarded, and her impartiality along with it, as she eyes black people for harsher punishment than most. The neutrality and fairness that are the bedrock of justice for the larger society are like quicksand beneath the feet of too many blacks. They must use extraordinary measures, including protests in the streets, appearances in the media, and appeals to local and national leaders to amplify their grievances, just to end up where most white citizens start. The folk of Ferguson, and millions more across the nation, have a difficult time getting the state that Obama represents to work on their behalf. In psychoanalytic terms, that is why Ferguson, and Baltimore after it, blew its id.

To his credit, Obama acknowledged that "a gulf of mistrust exists between local residents and law enforcement," and that "too many young men of color are left behind and seen only as objects of fear."[8] In one swift passage he spoke of "communities that feel left behind, who, as a consequence of tragic histories, often find themselves isolated, often find themselves without hope, without economic prospects," while their young men "end up in jail or in the criminal justice system" rather than "in a good job or in college." Later in his August 14 statement Obama briefly listed a set of "tends": black and Latino youth tend to face higher rates of school suspension, tend to have far more frequent interactions with the law, and may be subject to "different" trials and sentencing. Like the president himself, the language was careful and qualified, cautious, and perhaps a tad too clinical—a language that hardly captures the fiery realities that burn in black bodies and communities.

A Grammar of Impressions

What Obama said that day was true, but incomplete. Injustice is not simply a matter of perception, an instance in which blacks "feel" left behind or are subject to "different"—rather than inferior—brands of justice. The brute facts help explain why Ferguson combusted into shrieking anarchy: the decades of police aggression; the repeated killing of unarmed black people; the desperate poverty of black citizens; the entrenched bias in the criminal justice system and other institutions that are meant to help; the raging social inequality; the intended or inadvertent disenfranchisement of large swaths of the citizenry; and the dim prospects of upward mobility that grow bleaker by the day. A two-tiered system of justice operates for mainstream and minority communities. Blacks and other minorities often cannot get cops arrested when they are reasonably suspected of behaving unjustly, whether in Staten Island or Ferguson. They often cannot get their local municipalities to release autopsy reports and other pertinent information in a timely manner. And they often cannot make the local authorities treat their slaughtered loved ones like human beings, as they lie prostrate in the street for hours. Obama largely ignored these realities when he spoke from the White House the night of the Ferguson grand jury decision on November 24, 2014, about America as a nation of laws and said that we must respect the jury's conclusion, even if we do not agree with it, and make progress by working together—not by throwing bottles, smashing car windows, or using anger as an excuse to vandalize property or hurt anyone.[9]

The next day in Chicago, Obama doubled down on his indictment of "criminal acts" and declared, "I do not have any sympathy" for those who destroy "your own communities."[10] While he avoided saying so, it was clear that his remarks were directed at the black people who "looted" and "rioted" in Ferguson. But their criminal activity is the result of going unrecognized by the state for decades, a crime in itself. As for the plague of white cops killing unarmed black youth, the facts of which are tediously and sickeningly repetitive, and which impose a peculiar psychological tariff on black minds and exact a harmful toll on black bodies, the president was vague, halting, and sincerely noncommittal.

Instead Obama lauded the racial progress that he said he had witnessed "in my own life," substituting his body for our black bodies, his life for ours, and signaled again how his story of advancement was ours, suggesting, sadly, that the sum of our political fortunes in his presidency may be greater than the parts of our persistent suffering. As soon as Obama asserted that black folk are not delusional in saying they have big worries about the police, he reassured white America—he was not speaking to blacks, who need no such reassurance because most do not believe we have made it all up—that even though there are problems, that is not the norm. Even when he sidled up to the truth and nudged it gently—"these are real issues," the president acknowledged—he slipped back into an emotional blandness that underplayed the searing divide, saying there was "an impression that folks have" about unjust policing, and "there are issues in which the law too often feels as if it is being applied in discriminatory fashion."

Whose "impression" is it, though that word hardly captures the fierce facts of the case? Who feels it? Who is the subject? Who is the recipient of the action? Obama's treacherous balancing act between white and black, left and right—"there are good people on all sides of this debate, as well as in both Republican and Democratic parties," he said, which is true, but in what proportion he dared not say—posits a falsely equivalent relation that obscures the truth about who has held the power for the longest amount of time to make things the way they are. This is something, of course, that he can never admit, but which nevertheless leaves his words strained and turns an often eloquent word artist into a faltering, fumbling speaker. If language had hair and a face, Obama's grammar would be gray and weary. But his exasperated syntax is hardly a match for the fear and anxiety that black folk feel in the face of the police.

Policing the Black Body

It is nearly impossible to convey the fear that strikes at the heart of black Americans every time a cop car pulls up, emotions barely fathomable to whites, who do not generally view police as purveyors of urban terror. When I was seventeen, my older brother Anthony and I and a childhood friend were pulled over by four Detroit cops in an unmarked police vehicle. This was in the mid-seventies, in the shadow of the infamous Detroit Police Department task force called STRESS (Stop the Robberies, Enjoy Safe Streets), which was initiated after the 1967 riots. The unit lived up to its name and routinely targeted black folk. As we assumed the position against the car, I announced to one of the plainclothes officers that I was reaching into my back pocket to fish the car's registration from my wallet. He brought the butt of his gun sharply across my back and knocked me to the ground, promising, with a racial epithet, that he would put a bullet through my head if I moved again. When I rose to my feet, cowering, showing complete deference, the officer permitted me to pull out the registration. When the cops ran the tags, they concluded what we already knew: the car was not stolen and we were not thieves. They sent us on our way without a hint of an apology.

The lack of white empathy for black terror at the hands of the police came up at a meeting of black leaders with New York City police commissioner Bill Bratton that I attended in 2014 at the home of Citibank global banking head Ray McGuire and his wife, author Crystal McCrary. *CBS This Morning* co-anchor Gayle King reminded Bratton that "all black parents have had the conversation with their sons about being stopped by the police and how they have to manage the police's potential biases" against them. I pressed Bratton about the stop-and-frisk methods he had instituted twenty years earlier, and which continue today, despite their fueling racial disparities between "urban youth who get caught with roaches [the remains of a marijuana cigarette] in their pockets" and privileged college kids with "*Breaking Bad* meth labs in their dorm rooms."[11] The commissioner was far more receptive to our criticisms than the mayor who'd hired him to lead the city's police department in the mid-nineties.

My dustup in 2014 with former New York mayor Rudy Giuliani on national television tapped a deep vein of racially charged perception.

In a discussion on *Meet the Press* of Ferguson and its racial fallout, Giuliani steered the conversation down the path of a conservative shibboleth: that the real problem facing black communities is dying not at the hand of white cops but in the grip of black thugs.[12] He cited the statistic that 93 percent of black homicide victims are killed by black people; I argued that these murderers often go to jail, unlike the white cops who kill blacks with the backing of the government. What I did not have time to say was that 84 percent of white homicide victims are killed by white people, and yet no language of condemnation exists to frame a white-on-white malady that begs relief by violent policing. This does not mean that black folk are not weary of death ravaging their communities. I witnessed it personally as I sat in a Detroit courtroom twenty-five years ago during the trial of my younger brother Everett for second-degree murder, and though I believe to this day that he is innocent, I watched him convicted by an all-black jury and sentenced to prison for the rest of his life. But when the deaths of blacks somehow grant legitimacy to cops in the killing of often unarmed black people with impunity, the scales of justice are twice weighted against black interests. What is called for is active intervention on behalf of blacks and other citizens of color by their government — the same government that licenses cops to police black and brown communities.

In the face of it all, Obama did not offer public policies to address these ills, or the hope of politics based on true justice for all, but especially those hampered by race and class in their quest for that justice. Instead the president pivoted to the personal and suggested that his program to lift up black boys, My Brother's Keeper, would work with the Justice Department in "local communities to inculcate more trust, more confidence in the criminal justice system." But black youth do not need more trust; the justice system needs fundamental transformation. Obama, in a second statement on Ferguson on August 18, 2014, just days after his first, and well before the grand jury decision in November, proceeded to knock black youth while they were down by directing his law-and-order spiel against their already over-policed and under-protected bodies: "There are young black men that commit crime" who "need to be prosecuted because every community has an interest in public safety."[13] That is true, but tremendously tone-deaf in light of the denial of justice and the un-just criminalization of black people that led to a national crisis—for which Obama's most prominent answer was the recommendation of a social, not a political, program. Obama's defenders often claim that he is a president, not an activist, yet he sounded like one in his comments, and a bad one at that. In one rhetorical swoop, Obama leveraged the authority of the state against black youth, played to stereotypes of their criminality, offered responsibility lectures in place of public policy, maintained an emotional distance from the desperation of a group of Americans who happen to be his people, and offered them moral lessons instead of official action.

But these moral lessons, whether offered by Barack Obama or Rudy Giuliani, fall far short of the mark, substituting harsh reproof, false equivalence, and respectability politics in place of uplifting pol-icy. Many whites who point to blacks killing blacks are moved less by concern for black communities than by a desire to fend off criticism of unjust white cops. They earnestly believe that they are offering new ideas to black folk about the peril they foment in their own neighborhoods.

This brand of moralizing activism also found a champion in Bill Cosby, who for a decade had leveled moral charges against the black poor with an ugly intensity that was endorsed by white critics as tough love and by black journalists as homegrown conservatism. But Cosby's put-downs were more pernicious than that: his indict-ment of black women's lax morals and poor parenting skills was misogynistic. "Five, six children, same woman, eight, ten different husbands or whatever," Cosby fumed. "Pretty soon you're going to have to have DNA cards so you can tell who you're making love to. You don't know who dis is; might be your grandmother."[14] Cosby's Shakespearean fall from grace was attended by journalists who apol-ogized for having earlier failed to consider claims against the comic.

He was recast as a king who is more sinner than sinned against as the allegations of drugging and raping women piled up. But writers avoided mentioning their own sexist blinders that kept them from seeing how hateful Cosby was being toward black women long before he was accused of abusing mostly white women.

Cosby didn't invent the politics of respectability—the belief that good behavior and stern chiding will cure black ills, uplift black folk, and convince white people that blacks are human and worthy of respect. But he certainly gave it a vernacular swag that has since been polished by Barack Obama. The president has lectured blacks about their moral shortcomings before cheering audiences at college commencements and civil rights conventions. And yet his themes are shopworn and mix the innocuous with the insidious: pull your pants up, stop making racial excuses for failure, stop complaining about racism, turn off the television and the video games and study, do not feed your kids fried chicken for breakfast, be a good father.

As big a fan as he is of respectability politics, Obama is the most eloquent reminder that they do not work, that no matter how smart or sophisticated or upstanding one is, and no matter how much chastising black people pleases white ears, the suspicions about black identity persist. Despite his accomplishments and charisma, he is for millions the unalterable "other" of national life, the opposite of what they mean when they think of America. Barack Obama, like Michael Brown, is changed before our eyes into a monstrous thing that lacks humanity: a monkey, a cipher, a terrorist. One might expect the ultimate target of those who fear black otherness to have sympathy for their lesser targets, for people who have lesser standing and less protection, like the folk in Ferguson, in Cleveland, in New York, in Florida, in Baltimore, and all around the country, who cannot keep their unarmed children from being cut down in the street by callous cops who leave their slumped bodies to stiffen into rigor mortis in the presence of horrified onlookers.

Perhaps a measure of empathy lay behind Obama's sending Attorney General Eric Holder to Ferguson, though he should surely have gone himself, just as he went to Sandy Hook and to the areas struck by Hurricane Sandy. Sending Eric Holder to Ferguson was critical, but bringing Ferguson's blacks, and millions more like them, into Obama's presidential view, and into the folds of judicial fairness, would have been far more important. Obama, of course, is acutely aware of the tortured relations between law enforcement and black communities, which he has at times effectively recalled, as he did in his celebrated race speech in Philadelphia and in other pronouncements *before* he became president. Making such observations *as* president could help combat ignorance about the black plight in the criminal justice system, demonstrate healing compassion for black victims of police misconduct, and perhaps soften harsh attitudes toward black youth. This is a crucial role of the presidential bully pulpit — to speak with the authority of the office to tip the scales of moral fairness in favor of those who have been treated unjustly. Where Obama failed, his attorney general succeeded. Although Holder was not president, he certainly looked like one as he reached out to a bruised constituency, shook hands and kissed babies, promised fairness and the backing of the state to achieve justice, and reassured a demoralized population that their government cares for them. For all his lectures about responsibility to black audiences across the land, the president could have used a good dose of it himself.

Obama warned the unruly black elements in Ferguson that the nation is built on the rule of law. That is not entirely true. Obama's life, and his career, too, are the products of broken laws: his parents would have been committing a crime in many states at the time of their interracial union, and without Martin Luther King Jr. breaking what he deemed to be unjust laws, Obama would not be president today. Barack Obama is the ultimate paradox: the culmination of a churning assault on the realm of power that he now represents. No wonder he turns to his own body and story and life to narrate black bodies, black stories, and black lives. The problem is that the ordinary black person possesses neither Obama's protections against peril nor his triumphant trajectory that will continue long after he leaves office. And Obama's narrative does not answer a haunting question: If America can treat him as badly as it does, and he is as bright and affable as the best Americans, what will it do to the masses of Michael Browns in black communities? It should be remembered that Obama might have been Brown or one of the millions of black youth harassed by police and thrown into jail had he been busted for his youthful foray into drugs. That recognition sparked his My Brother's Keeper initiative. But that is only half the equation. What he must address somehow, with a fire that matches his condemnation of blacks murdering blacks, is my brother's killer, especially when the killer wears a badge and carries a gun on behalf of the country Obama embodies.

Fast and Slow Terror

The cruel reality is that nothing black people could say or do can change the minds of the white people who believe that black folks are a threat to them. They will neither love blacks nor leave them alone. Black people cannot be smart enough, good enough, or humble enough to please those who despise them, especially when they find legal cover for their animus behind a badge and a gun. Not even the election of a black president could unseat that stubborn fact. The murder of unarmed black motorist Walter Scott by white officer Michael Slager in South Carolina in April 2015 sheds light on political and social realities that surround similar cases of lethal force against black people. Americans have been forced to lower their expectations for racial justice and now measure racial progress in painfully minimal terms. The tragedy also brings into focus the optics of race—how black people are seen on camera and in history, revealing how black life is valued or degraded. And the spectacle of Scott's death highlights the fast terror that stalks black life even as it obscures the slow terror that blacks routinely confront in the Age of Obama.

There was great relief in black communities and beyond when

Slager was quickly arrested and charged with the murder of Scott. The deadliest moment of their brief encounter was caught on cell phone video: Slager drew his pistol and took aim at the fleeing Scott, unloading eight rounds and striking him dead. The video provided enough evidence to warrant an arrest, which is rare in police-involved shootings. Slager's jailing took place amidst the national outrage over a rash of unarmed black people like Michael Brown in Ferguson and Eric Garner in Staten Island dying at the hands of the police. During a recent seven-year stretch, a white police officer killed a black person nearly twice a week in America, underscoring the belief among blacks that they are targets of racial profiling and its violent twin, police brutality.[15] Many outside the black community think that the exercise of lethal force is warranted in most cases involving blacks and the cops.

Such clashing perceptions make it difficult to generate the broad consensus against police brutality in the Age of Obama that came to define the civil rights struggle against racial oppression in the sixties. The failure to find wide agreement has hampered racial progress in our criminal justice system, lowering the standards of racial justice, especially in contrast to the past. This has become brutally clear in the jarring juxtaposition of past and present. We have since 2013 experienced national celebrations of the triumphs of the civil rights movement—the March on Washington, the Civil Rights Act, the Voting Rights Act, and, in March 2015, the commemoration of Bloody Sunday in Selma. The crowning achievement of a black president in office as so many of these occasions are marked heightens the national appreciation of these jubilee celebrations. But the glory of the past runs up against the gory details of the present: an epidemic of black death at the hands of white police, flaws in a prosecutorial system that misrepresents the interests of black citizens, the failure of grand juries to indict cops in most police-involved shootings, and the vast overreach of a penal system in punishing people of color. Thus, when a white police officer is finally charged with murder, what should be seen as a minimal gesture is celebrated as a big victory because it took maximum effort to achieve. The floor of racial justice has been snatched from beneath the feet of black communities and turned precipitously into a ceiling—a harsh irony in Obama's America.

Our reverence for saints from the sixties underscores how we are addicted to the easy past rather than the hard present, though only a willful suppression of bitter facts can make us believe that anything about the racial past was easy—a narrative that President Obama has underscored. Its achievements were sealed in blood and cost the lives of some of the greatest witnesses for political transformation the nation has ever had. But Americans are bad at taking in race in real time; we prefer rose-tinted lenses for watching slow-motion replays in which we control the narrative and downplay our complicity in the horrors of our history. Unfortunately, President Obama's racial procrastination has only exacerbated this tendency.

The racial present is messy, unresolved; it thumbs its nose at stories that promote bland racial optimism about how far we have come. Every black body that suffers unjustly at the hands of a cop chips away at racial triumphalism. Racial optimism and racial triumphalism make it more difficult to organize resistance and gain white allies. There have been many white participants in the series of protests across the nation against police brutality who remind anyone within earshot, in their familiar chant, that "Black Lives Matter." These actions echo a past when blacks and their allies forced the nation to grapple with its racist legacy through acts of civil disobedience that were harshly criticized and resisted.

Many Americans in the past finally conceded the legitimacy of black struggle because its leaders brilliantly staged their protests for the world to see. White citizens struggled to digest their meals in peace as scenes from Selma's bloodbath flashed on their television screens. It was more difficult to write off Negro complaints as gestures of self-pity when the fangs of police dogs tore at the flesh of women and children on the evening news. But that, and Americans, and their media, have all changed. Massive black marches have diminished, American guilt and compassion have severely flagged, and American television has been radically transformed: a thousand or more stations compete for our attention, and the rise of the Internet and social media has challenged television in supplying the unifying fiction of American identity and citizenship.

The fractured media landscape has led to the proliferation of images, perception overload, and a vying for digital attention. The demand for spectacle merges with the desire to capture more eyeballs on television, computer, and smartphone screens. This visual barrage and optic glut make it difficult to command the unified national consciousness in the same way as when there were only three networks in play. We are reduced to forging workable rather than wide consensus, more modest goals for justice, and shorter-term alliances with potentially increased numbers of allies, for digital ties are not necessarily those that bind, even if they point to larger landscapes, longer timelines, and deeper truths. The digital can in fact be the handmaiden of the historical when rightly used. President Obama brilliantly proved it when he transformed the American political campaign with his unprecedented success in fund-raising and message-sharing on the Internet.

But all of this seeing and overseeing in contemporary visual culture does not solve the problem of how black folk have been historically viewed in a negative cultural light. Even the sight of a black president whose image is posted daily in cyberspace, plastered on print newspapers across the nation, and televised around the world cannot dislodge the set of images that fix black life in the national and global glare. Tragically, the negative thinking about black life has survived media transformations and Obama's rise to power. Another way of saying this is that the content of black identity has survived a change in format and presentation. New media, besides breaking barriers so people of color can speak up, has also provided the culture with more ways to stereotype, more ways to be suspicious and hateful.

The optics of race are tricky: while contemporary media and devices allow us to see more—including images of brilliant and beautiful black people occupying the White House and representing the nation's family values—they do not necessarily allow us to see more deeply. That millennials see race the same way Generation Xers and baby boomers do testifies to a troubling racial continuity. Moreover, stereotypical representations of blackness—some authored by black hands and disseminated as reality TV—are accepted as normative. Yet problems arise when images of blackness contradict a received racial script. That is why it was easier to believe that the video footage of Michael Brown in Ferguson stealing cigarillos more accurately communicated his character as a "thug" than to believe that the last gasps of Eric Garner were the pleas of an unjustly assaulted man. The Michael Brown video reinforced the belief that black males are inherently criminal; the Garner video, despite what we saw, contradicted the script that says even an unarmed black man begging to breathe cannot be believed—that says he is literally lying as he lies on the ground dying. We cannot believe what we see because it contradicts what we have seen and been led to believe. What we see is not simply determined by what we perceive with our eyes; instead, sight registers the cumulative impact of what we learn and what we think we know.

Thus a history of how blacks have been seen is recapitulated each time a new video surfaces of black people being poorly treated by the police—from Rodney King to Walter Scott to Dajerria Becton in McKinney, Texas. But what we see with our eyes is often contradicted by what we see with our collective sight in a culture that has taught us to understand blackness in an especially malevolent fashion. Thus, before the video of his encounter with Scott emerged, Slager relied on a script that many white cops have used, including Darren Wilson in Ferguson: I was afraid for my life; the black man reached for my weapon to harm me; I had to defend myself with lethal force. Those police scripts derive from a larger pool of stories about black people as dangers and threats, and thus these cops' stories make sense to the majority of white Americans because they have fed on a common diet of black perception. Police reports sync up with images derived from our culture.

Also, what looks obvious to black folk—that they are under siege—seems to shift when white eyes land on black subjects, as people and as issues. Whites and blacks see from two different perspectives shaped by history. For instance, black and white Americans view the presidency of our first black commander in chief in widely varying terms. What blacks see is sometimes not viewed as rational or real, or worthy of respect; it does not count as sufficient evidence to prove a case of abuse or injustice. This is why President Obama repeats, in the wake of police shootings of unarmed victims, that black people are not making up their perceptions of injustice. The demand for proof of what they believe is not foolproof: even when video evidence emerges, it is not seen by many whites as incontrovertible or even persuasive of the case made by blacks that mistreatment abounds. There is a racial Rorschach test going on: we see the same image, but we sometimes do not see the same reality, or the same truth it reveals. This proves that seeing involves more than sight; it involves "sites," too, of past realities packaged in ready-made images of, for example, black pathology, or deserved poverty.

Rodney King was brutally beaten by police, but a jury acquitted his abusers; Eric Garner pleaded for his life, but a grand jury failed to indict. It is not just what is seen; it is what scenes of race replay in our heads. Black frustration mounts when blacks have what they think is clear evidence of police misconduct, and the failure to appreciate black life is reinforced when there is a rejection of what stands as proof that their lives do not matter the same way as white lives. They cannot matter the same way because they cannot be seen the same way. To make matters worse, the fictional images of blacks held in many minds are taken as literal, while the images from real-life cameras fail to convince whites of what blacks see: that black lives do not matter as much. The smartphone has turned the spectator into a participant, permitting her to record and change history. Given black people's ready experimentation on the cutting edge of beepers, pagers, and cell phones, that shift seems to favor them. In the case of Walter Scott, a police shooting caught by cell phone, his police assailant was charged with murder by a grand jury. But often the electronic evidence will not relay truth back, because the broader context can never be underestimated or dismissed or ignored. This is why Obama's importance as a public historian and interpreter of racial experience can hardly be disputed.

This failure to be taken seriously, or to be seen in the right contexts, and to be seen as human, is part of the trauma of black existence, one that reinforces an often submerged truth: the lived experience of race often feels like terror for black folk, whether fast or slow. Few metaphors and tropes more adequately capture what it means to be black and afraid of random and arbitrary forms of violence than the single word "terror." If the American people now believe they are subject to assault from forces in the Middle East for no other reason than that they are American, that comes close to what it means to be black: for no other reason than their identities, blacks are pro-filed, abused, dismissed, disbelieved, set aside—literally killed and un-mattered. Black people know what it means to feel insecure in one's home, unprotected by one's government; no space is safe or adequate to prevent the plague of assault just because one is black.

The recording of Walter Scott's death is so terrifying because it could be any black—the real fear that terror seeks to impart. In most cases in the past, and likely in the future, there were, and will be, no cameras to vouch for blacks, no incontrovertible evidence that blacks were assailed; no matter how much education blacks possess, how much money they have in the bank, how many late-model cars they drive, how well behaved they are, how articulate they become, they may, at any moment, be gunned down or feel a baton beating them, a Taser electrifying them, a bullet penetrating their flesh—all because they are black, and therefore seem likely to commit, or to have committed, a crime. Ironically, blacks are seen as necessary sac-rifices for the safety of white society; they are viewed as scapegoats, or perhaps collateral damage, in the white war against the terror of black criminality.

The terror that black people experience is of two varieties. *Slow terror* is masked but malignant; it stalks black people in denied opportunities that others take for granted. Slow terror seeps into every nook and cranny of black existence: black boys and girls being expelled from school at higher rates than their white peers; black men and women being harassed by unjust fines from local municipalities; having billions of dollars of their wealth drained off by shady financial instruments sold to blacks during the mortgage crisis; and being imprisoned out of proportion to their percentage of the population. President Obama has referred to this kind of terror as a "slow rolling crisis." *Fast terror* is more dynamic, more explicitly lethal, more grossly evident. It is the spectacle of black death in public displays of vengeance and violence directed against defenseless black bodies. Shootings like that of Scott traumatize blacks, too, because they conjure the historic legacy of racial terror: lynching, castration, and drowning. The black body was not safe then, and blacks today do not feel safe, or accepted, or wanted, or desired.

The last moments of Scott's life, caught on video and widely watched, are classic fast terror. The video is sickening because it captures the breathtaking indifference to moral consequence that seems to grip Slager as he fires at an unarmed black man in broad daylight. A frozen frame from the video shows a police officer, gun drawn, in pursuit. Fifty years earlier, a lawman in pursuit pulled his gun and shot dead the Selma protester Jimmie Lee Jackson, whom Martin Luther King called a "martyred hero of a holy crusade for freedom and human dignity."[16] The failure to be seen as human unites black people across time in a fellowship of fear as black people share black terror, at both speeds, in common. The way we see race plays a role in these terrors: fast terror is often seen and serves as a warning; slow terror is often not seen and reinforces the invisibility of black suffering. Fast terror scares black people; slow terror scars them.

Black, White, Blue, and Gray

The way fast and slow terror occasionally entangle in menacing indivisibility played out in Baltimore in the aftermath of the funeral for Freddie Gray, the young black man who died from a spinal cord injury in April 2015 while in police custody. Gray's arrest, like Scott's murder, was captured on cell phone camera video as he was dragged into a police wagon by several officers. The six police officers involved in arresting and transporting Gray were quickly charged with crimes ranging from manslaughter and illegal arrest to second-degree depraved heart murder. It was a rare display of prosecutorial vigilance on behalf of black victims of alleged police misconduct. A grand jury largely agreed with Baltimore City state's attorney Marilyn Mosby when it later indicted the officers on most of the charges Mosby brought; though it dropped charges of illegal imprisonment and false arrest, the grand jury added a charge of reckless endangerment against all the officers involved.[17]

Before any of the charges were filed, the beleaguered community gathered, metaphorically, in the spacious sanctuary of Baltimore's New Shiloh Baptist Church, at the funeral of Freddie Gray, as the familiar weight of grief descended over participants, not just for Gray but for the countless Freddie Grays across black America who die, unarmed, at the hands of the police. A few hours after the Gray family laid their son to rest, the city's black anguish burst into flame as cars were burned and young people hurled rocks at cops.

A predictable question trailed closely behind their actions, a question that always reappears like the ghost of riots past: Why are they destroying their own neighborhoods and setting their futures on fire? The question feels helpless, sometimes cynical, but it is exactly the right question. It should be asked, however, not in anger but with compassionate curiosity. Because the truth is as ugly as the facts that fuel riots: without a brick tossed or a building burned—the dramatic response to the fast terror of Gray's death—the nation hardly confronts the hopelessness, the slow terror, of the future for these young people, a point that President Obama underscored in the smoldering aftermath of Baltimore's enflamed grief.

Sadder still is that the social neglect that sparked the carnage had been largely overlooked by the powers that be. The unemployment rate in the community where Gray lived is over 50 percent; the high school student absence rate hovers around 49.3 percent; and life expectancy tops out at 68.8 years, according to analyses by prison reform nonprofits.[18] These statistics are a small part of the portrait of radical inequality and slow terror that blanket poor black Baltimore. It is no wonder that black Baltimore erupted in social fury. As Martin Luther King Jr. announced in the wake of the Watts riots fifty years ago, "A riot is the language of the unheard."[19] Judging by the actions in Baltimore, thousands were not being heard. The stale repetition of black death at the hands of the police led to burning rage, a rage that would, sadly, lead elected officials, including President Obama, to label participants in the mayhem "thugs"—while no such label, nor the word "riot," left their mouths to describe the 9 deaths, 170 arrests, and destruction in the wake of a melee between dueling biker gangs in Waco, Texas, a month after the Baltimore uprising.[20]

Perhaps a basketball analogy can explain the urban rebellion. Often on the court, a player commits an offense—say, hitting an opposing player in the ribs—without being spotted by the referee. Then, when the offended player strikes back, he is the one hit with a foul. The black youth who took to the streets have been hit with the slow terror of so many unacknowledged assaults—from racial profiling to poor schooling—that their violent responses are frequently viewed through a haze of social stigma that penalizes them without regard for context.

Jesse Jackson—who helped to eulogize Gray, arguing at his funeral service that the young man was now "more than a citizen" and had become "a martyr"—spoke of that context, and reflected on the conditions of slow terror that plague black communities and leave black youth destitute. "Our boys are the least educated, the most profiled, and the most jailed, do the most prison labor, [are] the most unemployed and have the shortest life expectancy," Jackson lamented. He acknowledged the chaotic consequences of social injustice in black communities. "When there is darkness there will be crime and behavioral issues," the seasoned minister observed. "It is easier to fight the victim rather than the source of the darkness." Jackson also admitted that the presence of new technology advanced the quest for justice. "In some sense what makes the difference today is his innocence, and the presence of a camera," he said of Gray. "If he had been in a gun shootout with the police, or if he had been killed in a drug bust, or caught in some compromising position hurting someone[,] it would have been easier to dismiss his killing." Jackson touched on the furious tensions between the police and black communities that make a mockery of any sense of security and justice. "The Baltimore police became the pallbearers of an alive man and turned the paddy wagon into a tombstone," Jackson charged. "We are here because we all feel threatened. All of our sons are at risk. Their number has just not popped up yet. There is too much killing, too much hatred, and too much fear."[21]

All Black Lives Matter

It is not just black sons who are threatened by or at risk of police brutality but black daughters and mothers as well. The horrifying spectacle, caught on cell phone video, of fifteen-year-old bikini-clad Dajerria Becton being violently manhandled by a white police officer, Eric Casebolt, at a pool party in McKinney, Texas, is a graphic example. After an earlier fight between white parents and black teens—the white parents allegedly made racist statements, including telling the youth to "return to Section 8 housing," and slapped a black female party participant—police were called to the scene to calm any further disturbance. The video shows Police Corporal Casebolt cursing and screaming at Becton and her friends, ordering them to leave the area and later warning them not to "keep standing there running your mouths." If the angry black male is a stock character in racist mythology, the sassy, loudmouthed, back-talking black woman is another staple. As Dajerria obeys Casebolt's command, the officer inexplicably yells at her, chases her down, grabs her arm, and drags her from the pavement onto the grass.[22]

Dajerria spoke later in a television interview, saying, "[Casebolt] told me to keep walking and I kept walking and then I'm guessing he thought we were saying rude stuff to him. He grabbed me and he like twisted my arm on the back of my back and he shoved me in the grass. He started pulling the back of my braids and I was like telling him that he can get off of me because my back was hurting really bad."[23] The video shows Casebolt violently slamming Dajerria to the ground and forcefully planting his knee in her back as she cries out for her mother while her friends plead with Casebolt to stop.

Two of the teens witnessing Casebolt's violent outburst move toward the officer; he suddenly ceases pressing Dajerria to the ground and jumps up, pulls out his revolver, and points it at the youths. After fellow officers restrain him, Casebolt returns to Dajerria and places her in handcuffs. No charges were filed against Dajerria, and Casebolt resigned a few days after the event, but the message the incident sent was no less chilling: black youth peaceably assembled in their suburban community are nevertheless the cause of undue suspicion and unwarranted harassment from both the police and the broader community. They are often subjected to a form of violent state response that was not displayed in April 2014 when white Nevada rancher Cliven Bundy and fellow protesters brandished guns in a dispute with the federal government over unpaid grazing fees — or with the mostly white bikers in Waco who murdered nine members of rival gangs.

Black girls and women of all sexual orientations have been erased from portrayals of both the slow and fast terror that black people endure, despite the enormous work they have done to amplify the voices of the unjustly aggrieved, symbolized in the important work of Kimberlé Crenshaw, Columbia law professor and co-author with Andrea J. Ritchie of the report "Say Her Name: Resisting Police Brutality Against Black Women,"[24] and the Black Lives Matter movement begun by two black queer women, Alicia Garza and Patrisse Cullors, and by Opal Tometti. The Say Her Name protests in 2015 in several cities across America were fashioned by activists to highlight the countless black women who have been harassed, harmed, and even murdered by the police.

Perhaps no other figure has symbolized the complicated status of black women in America, and the micro-aggressions and the slow and fast terror to which they are subjected, more prominently than First Lady Michelle Obama. Many observers initially cast her as the angry black woman, while others framed her as the castrating shrew to her famous husband, Barack—reinforcing black female stereotypes that found sociological sway in the Moynihan Report on the black family. Michelle Obama spoke about these visions of her, and the pain they evoked, in a remarkable passage from a commencement address she delivered in 2015 at Tuskegee University:

> Back when my husband first started campaigning for President, folks had all sorts of questions of me: What kind of First Lady would I be? What kinds of issues would I take on? Would I be more like Laura Bush, or Hillary Clinton, or Nancy Reagan? And the truth is, those same questions would have been posed to any candidate's spouse. That's just the way the process works. But, as potentially the first African American First Lady, I was also the focus of another set of questions and speculations; conversations sometimes rooted in the fears and misperceptions of others. Was I too loud, or too angry, or too emasculating? Or was I too soft, too much of a mom, not enough of a career woman?
>
> Then there was the first time I was on a magazine cover—it was a cartoon drawing of me with a huge Afro and machine gun.

Now, yeah, it was satire, but if I'm really being honest, it knocked me back a bit. It made me wonder, just how are people seeing me.

Or you might remember the on-stage celebratory fist bump between me and my husband after a primary win that was referred to as a "terrorist fist jab." And over the years, folks have used plenty of interesting words to describe me. One said I exhibited "a little bit of uppity-ism." Another noted that I was one of my husband's "cronies of color." Cable news once charmingly referred to me as "Obama's Baby Mama."

And of course, Barack has endured his fair share of insults and slights. Even today, there are still folks questioning his citizenship.

And all of this used to really get to me. Back in those days, I had a lot of sleepless nights, worrying about what people thought of me, wondering if I might be hurting my husband's chances of winning his election, fearing how my girls would feel if they found out what some people were saying about their mom.[25]

It is her racial candor that distinguishes Michelle from her husband—a candor, in all fairness, denied to Barack because of the position he holds—and yet it underlines how much the nation misses when Barack Obama fails to do what he might reasonably be expected to do as an American president: tell the truth about race and make public policy yield to the democratic demand for just governance. Michelle Obama's achingly honest remarks drew predictable criticism from right-wing quarters, but her forthright expression of the existential angst unleashed by a society not yet at peace with its racial past is precisely the sort of testimony the nation needs to hear from voices that matter.[26] (The fact that, according to a former Secret Service agent, it was Michelle who privately pushed her husband to "side with blacks in racial controversies" only burnishes her reputation for being Barack's black conscience.)[27]

President Obama seemed to learn from his wife's candor in the ongoing effort to counter the plague of lethal policing that has engulfed poor black communities since Ferguson. In an April 2015 joint press conference with the Japanese prime minister, Shinzo Abe, Obama was as direct as he had ever been in finally acknowledging, "There's some police who aren't doing the right thing."[28] Obama argued that since "Ferguson . . . we have seen too many instances of what appears to be police officers interacting with individuals—primarily African American, often poor—in ways that have raised troubling questions." Obama acknowledged the legitimacy of civil rights leaders' and black parents' thinking of the lethal interactions as a crisis—though in his words, it was a "slow-rolling crisis" that has been "going on for a long time . . . [W]e shouldn't pretend that it's new." What was new, Obama suggested, was the broader public's awareness, because of social media and cell phone video, of a catastrophe of police brutality that black folk have endured for decades.

Obama avoided strategic inadvertence, and even resisted for a spell the call of the noble implicit, and argued that there were deeply rooted systemic issues that had to be confronted beyond a focus on cops. These include "impoverished communities that have been stripped away of opportunity, where children are born into abject poverty" to parents plagued by substance abuse problems, low levels of education, and high levels of incarceration, producing another generation whose "kids end up in jail or dead" rather than going to college. Obama spoke of poor communities with "no investment," lost manufacturing, and the rise of a drug industry as the primary employer. "If we think that we're just going to send the police to do the dirty work of containing the problems that arise there without as a nation and as a society saying what can we do to change these communities, to help lift up those communities and give those kids opportunity, then we're not going to solve this problem." Obama warned that the failure to address these issues adequately means "we'll go through the same cycles of periodic conflicts between the police and communities and the occasional riots in the streets, and everybody will feign concern until it goes away, and then we go about our business as usual."

Obama made a passionate plea for society's responsibility to address the crisis:

> If we are serious about solving this problem, then we're going to not only have to help the police, we're going to have to think about what we can do—the rest of us—to make sure that we're providing early education to these kids; to make sure that we're reforming our criminal justice system so it's not just a pipeline from schools to prisons; so that we're not rendering men in these communities unemployable because of a felony record for a nonviolent drug offense; that we're making investments so that they can get the training they need to find jobs . . .
>
> [I]f we really want to solve the problem, if our society really wanted to solve the problem, we could. It's just it would require everybody saying this is important, this is significant—and that we don't just pay attention to those communities when a CVS burns, and we don't just pay attention when a young man gets shot or has his spine snapped. We're paying attention all the time because we consider those kids our kids, and we think they're important. And they shouldn't be living in poverty and violence.

This is Obama at his best: analyzing the broad sweep of social distress and accepting responsibility to address problems bigger than individual will and beyond the sway of personal accountability. In doing so, Obama serves the interests of a besieged black constituency, and therefore the interests of the country, far better than when he ignores race, denies white responsibility, or criticizes black culture. While all citizens have the responsibility to contribute to the common good, not all citizens are equally able to attend to their welfare and fend off suffering, especially when the state has had a big hand in creating their plight. Race has fatefully shaped the destinies of the nation's black citizens. The greatest American presidents have memorably wrestled with the destructive legacies of racism to make the nation a true democracy for all its citizens.

In our time, that includes the black girls and women whose needs and challenges Obama finally addressed in his 2015 speech to the annual gala of the Congressional Black Caucus, and in a White House Summit on women and girls of color later that year. It also embraces the youth of the Black Lives Matter movement, whose ideas and activists Obama validated in an October 2015 White House forum on criminal justice.

"I think the reason that the organizers used the phrase 'Black Lives Matter' was not because they were suggesting nobody else's lives matter," Obama said. "Rather, what they were suggesting was there is a special problem that's happening in the African American community that's not happening in other communities.

"And that is a legitimate issue that we've got to address."

Occurring late in his presidency, these statements—along with his declaration in October 2015 to the International Association of Chiefs of Police that earlier in his life he sometimes got tickets he didn't deserve, and "that when you aggregate all the cases . . . there's some racial bias in the system"—prove that Obama has struggled to find his voice in the face of race; as the nation's first black president, he has been as much a victim of our poisonous racial compact as he has been the arbiter of the state's response to racial tragedy and its commitment to racial justice. Finding the nerve to tell the truth about race could only deepen his considerable legacy.

7

GOING *BULWORTH*

Black Truth and White Terror in the Age of Obama

IN A 2013 *NEW YORK TIMES* ARTICLE DISCUSSING PRESIDENT OBAMA'S frustrations with obstruction in Congress, policy problems in the Middle East, and accusations of the IRS targeting Tea Party members, all in the tender days of his second term, the commander in chief chafed at the limits of the presidency. Privately he wished to "go *Bulworth,*" referring to a 1998 Warren Beatty film about a senator who spurns caution and bluntly speaks his mind. As the *Times* article observed, while Beatty's character "had neither the power nor the platform of a president, the metaphor highlights Mr. Obama's desire to be liberated from what he sees as the hindrances on him."[1] Several journalists noted the president's exasperation and projected what liberation might sound like, yet only a few considered how, or if, it meant that Obama might tell the truth about race.[2] Obama's legendary discipline on the subject had even shaped the fantasies of writers wondering what he might talk about if he were free to speak.

It is curious that race was barely mentioned in these discussions, since it occupies the heart of Beatty's kinetic morality tale about the political and personal cost of telling the truth. Beatty found the rhymes and reason of gangsta rap an irresistible medium to broadcast brutal truths, smashing black respectability and white indifference in a single cinematic swoop.[3] Perhaps Obama's yen for a *Bulworth* reckoning is a sign of his racial unconscious coming clean, or at least getting loud and conscious. Obama's attraction to *Bulworth* inspires a reckoning with the film's message about race and black culture. Obama's fantasy of free speech, however, need not have breathed through a fictional character; in his own administration, the first black attorney general managed to tell great truths about race while taking care of his broader responsibilities. Eric Holder's outlook and approach to race may be more effective than Obama's more cautious tack in countering the racial terror that targets black America — as in the brutal murders of nine black parishioners in a Charleston, South Carolina, church in June 2015 by white supremacist Dylann Roof. The massacre provoked in Obama a surprising willingness, no matter how late in his term, or how briefly, to speak bluntly about race, leading one to ask if Obama finally did achieve a fleeting *Bulworth* moment.

White Skin, Black Masks

Warren Beatty's *Bulworth* offers a frenzied and rare mash-up of two of Barack Obama's passionate constituencies: liberal politics and hip hop. Beatty grasped the cinematic force of rap music and its brilliant deconstruction of the thuggish hypocrisy of politics. Hip hop's most provocative images have always come camera-ready. The music video, which hip hop grew up with, helped to forge a seamless flow of images from ear to eye, extending the music's aesthetic function into full-length features. From low-tech Instamatic snapshots of an evolving art form (1985's *Krush Groove*) to the zany attempt of artless talents to get paid (1993's *CB4*) — and from the erudite quest of three

Afro-Bohemians to find '90s hip hop bliss amidst a twisted crime scheme (2015's *Dope*) to the epic gangsta neorealism of artists facing terror on the streets and from the police (2015's *Straight Outta Compton*)—the hip hop film has best succeeded when it captures the moral complexities of the music it emulates. This fact also made Rusty Cundieff's 1994 film *Fear of a Black Hat* a dead-on parody of hardcore rap: it caricatures the desire of rappers to be thugs in order to satisfy fans whose desire for thugs is driven by the stories of hardcore rap.

Hardcore rap has been swollen with gangster images plucked from the cinematic imaginations of directors Francis Ford Coppola and Martin Scorsese, its hoods caramelized knockoffs of characters immortalized by actors Robert De Niro and Al Pacino. It matters very little that black gangster films came first. After all, seventies blaxploitation flicks like *Black Caesar, The Mack,* and *Superfly* amply attest to the violent fantasies that crammed the urban imagination of many of that era's artists. What does matter is that hardcore hip hop has reinvented the manner in which black gangsterism is conceived. Gangsta rap has, in a sense, birthed its predecessors by helping us understand why they were interesting and how they made sense. In the process, it rendered plausible what was only dimly and sporadically acknowledged about gangsterism in civil circles: that it pervades American culture, that its offenses are counted more heavily when its complexion is dark, and that its racial subtexts provide an unflattering mirror of national self-examination. In that way, the hip hop that President Obama likes, especially the music of Jay Z, whose lyrics teem with tales of hustling, may provide the soundtrack to his unspoken desire to tell grittier truths about politics and race.

It may also explain the appeal to Obama of *Bulworth,* which, like the president's biography, joins black and white elements in complicated, intriguing fashion. The film's creation was certainly paradoxical: a wealthy, legendary white actor and lionized auteur, perhaps one of the last great movie stars, turned his attention to the ugly and inconvenient specificities of racial and economic inequality.

"I paid my dues in politics," Beatty told me in an interview. "So I knew at some point I would do a picture about contemporary politics." Beatty admits, however, that "it was very hard to do it straight because [politics] is so ridiculous." *Bulworth* is a political satire about a despondent white politician of the same name, played by Beatty, who, in a moment of extreme depression, contracts to have himself murdered so that his family can collect his multimillion-dollar insurance policy. Arranging to be killed by the end of the weekend, Senator Jay Billington Bulworth feels free to do what under normal circumstances would spell sure political death: tell the truth, even if it happens to be politically incorrect and, on the surface at least, highly offensive. (Given the bubble presidents live in, it's understandable why the fantasy of plain talk would appeal to Obama).

For instance, at a black church rally, when asked by a black woman why he had not shown up since being elected, Bulworth confesses that he really did not care about blacks beyond getting their votes. Adding insult to injury, he claims that until black folk stop "drinking malt liquor" and "eating chicken wings," and until they cease supporting "a running back who stabs his wife" (a reference to O. J. Simpson), they will never get rid of politicians like him. And at a Beverly Hills fund-raiser attended by rich Jews in the film industry, Bulworth offends them by assailing their moral hypocrisy in making sleazy movies. Insulting them even further by offering mocking appeasement, Bulworth, referring to a speech written for the occasion by his chief political aide, Dennis Murphy (played by Oliver Platt), says, "Murphy probably put something bad about Farrakhan in here for you."[4]

If one is a literalist, one might think Beatty, through his Bulworth character, is playing to stereotype and bigotry. One might also conclude that he's in sync with the president, at least in speaking of black culture, since Obama has evinced a disturbing inclination to embrace stereotype in his black respectability sermons. But if one is alert to the satirical, it is clear that the function of such lines in *Bulworth* is to draw a contrast to the usual duplicity of public figures who mask their beliefs with impenetrable doublespeak. At the black church rally, Bulworth meets two "b-girls" from the hood who volunteer to help his campaign. Bulworth is unfazed until he lays eyes on Nina (played by Halle Berry), a beautiful woman whom the married senator immediately lusts for. Bulworth's meeting with Nina proves fateful: it gives him a reason to live; it gives him a renewed sense of his political vocation; and it introduces him to the pleasures (whizzing around with the "b-girls" in his limo, partying until dawn, the intimate enjoyment of black slang, the freedom to celebrate life) and the pain (the plague of police brutality, the vagaries of drug addiction, the menace of black male hatred, returned in full by the senator) of urban black life. Although it debuted nearly twenty years ago, *Bulworth* is, in some ways, as relevant now as when it first appeared because of the racial and political forces the film exposes, forces that have again claimed center stage in the Age of Obama.

After falling for Nina, Bulworth adopts a hip hop style of dress and speech to address his bewildered public and television audiences. "It's just more fun if out of my middle-aged white body comes a young black protesting person," Beatty told me, "even if he's behaving in an adolescent way." While Beatty disavows using "movies as propaganda," he acknowledges, "There are times when a social agenda is compatible with a movie."

Rhyming on television and before white audiences, Bulworth sounds at times like Jesse Jackson on crack: a hyper-verbal politician turned public moralist whose cadences both mock and reinforce the culture he has appropriated and been influenced by. In the hands of a lesser artist, this might all collapse into an unintended parody of a white man failing to make sophisticated use of black culture because he fails to understand the difference between satire and minstrelsy. But Beatty gets it just right. As *Bulworth*'s star, director, and cowriter, Beatty uses gangsta rap's erotically charged violence and vulgar speech, both literally and metaphorically, to reveal the corruption of electoral politics.

"Clearly what I think is obscene is the disparity of wealth and inequality in the country," Beatty stresses to me. "I don't think words like 'fuck,' 'motherfuck,' 'cocksucker' are obscene. They are attention-getting words. The real obscenity black folks are living with is trying to believe a motherfucking word that Democrats and Republicans say." In *Bulworth*, Beatty is not only flipping the script. He is unabashedly embracing an art form that has been scorned by white politicians and by the black bourgeoisie. In colorful and comedic terms, *Bulworth* shows how the social rituals and cultural conventions of gangstas and politicians are driven by the same goals: getting paid, getting pleasure, and getting props. (Obama certainly agrees with portions of *Bulworth*'s argument, especially about the corrupting influence of big money in politics, and has said so, but cannot speak in the blunt terms the film adopts.)[5]

Equally effective in the film's effort to tell the black truth is Nina, whose character crushes stereotypes and supplies the film's intellectual and ideological heart. While she is being romantically pursued by Bulworth, Nina drops a brilliant explanation of how economic forces have devastated inner-city communities, robbing them of the hopefulness that produces strong black leadership. Nina's mini-speech—coupled with a fiery morality tale from drug dealer L.D. (played by Don Cheadle) about the infusion of crack cocaine and other drugs into black communities, and the informal political economies that are supported by white negligence and black capitulation—gives *Bulworth* an added political function: it legitimates progressive and grassroots viewpoints about the scourge of drugs and criminality in black communities. Obama the community organizer may have sung a similar melody, but Obama the president could scarcely afford to hum that tune. There is little doubt that, because Beatty is a white male and a beloved American icon, he is able to present what have been considered paranoid conspiracy theories of the black fringe as reasonable expressions of political common sense.

Bulworth's unsparing indictment of the unjust concentration of capital and political resources in the hands of the very rich—especially, in this case, in the coffers of insurance companies—joins together the issues of race and wealth more successfully than most presidents are capable of doing. In fact, Senator Bulworth's moral and political transformation at the hands of black culture — specifically, at the hands of hip hop culture, and gangsta rap at that, allegedly the most antipolitical of hip hop's genres — gives him incentive to avoid the assassin's treachery he has paid for. But what *Bulworth* does best is embrace black culture while telling the truth about America.

Life Imitating Art?

One wonders what it might look or sound like if Obama had the opportunity, or perhaps the courage, to adroitly, and discreetly, sample the insights of hip hop, or Beatty's film, in speaking to the masses. The spirit of *Bulworth* seemed to have briefly taken hold of Obama in June 2015 as he addressed the question of same-sex marriage in the East Room of the White House while a pro-immigration advocate interrupted his speech. Obama has in the past displayed remarkable forbearance with hecklers, but he was having none of it that day.

"Hold on a sec," Obama said to the heckler, later identified as Jennicet Gutiérrez, a transgender woman who called on the president to release all LGBTQ immigrants from detention and to stop all deportations.[6]

"OK. You know what. Nah, nah, nah, nah, nah, nah. No, no, no, no. Hey, listen, you're in my house," Obama said, pointing a finger at Gutiérrez as the audience cheered. "You're not going to get a good response from me by interrupting me like this."

A chorus of boos went up.

"No. Shame on you, you shouldn't be doing this."

A chant of "Obama" echoed in the room as the president directed that Gutiérrez be removed from the audience. After she was escorted out, Obama got even feistier.

"As a general rule, I am just fine with a few hecklers," Obama said. "But not when I'm up in the house!" His casual resort to black vernacular was greeted with cheers and laughter. "My attitude is if you're eating the hors d'oeuvres . . ."

Obama wheeled around to look at Vice President Biden, who was smiling and nodding along with Obama while gripping the president's shoulder.

"You know what I'm sayin'?" Obama asked.

"I know what you're saying," Biden shot back in a brief call-and-response with the president.

"And drinkin' the booze," Obama continued, and then humorously offered a colloquial self-affirmation in the style of the black church: "I know that's right!"

Earlier in the year, Obama had eagerly embraced the mirthful pose of *Bulworth* to deliver some tart truths at the 2015 White House Correspondents' Association Dinner. In his speech Obama alluded to his favorite sport, basketball, and acknowledged being in "the fourth quarter of my presidency," feeling "more loose and relaxed than ever before" and "determined to make the most of every moment I have left."[7] Obama quipped that after the midterm elections his advisers had asked if he had a bucket list. "Well, I have something that rhymes with bucket list," the president deadpanned as the audience erupted in laughter. "Take executive action on immigration? Bucket. New climate regulations? Bucket. It's the right thing to do."

Obama also drew big laughs when Keegan-Michael Key, one half of a comic duo, joined him at the podium as "Luther"—the character he played on the hit Comedy Central show *Key & Peele*—who humorously acts out Obama's (seldom otherwise visible) angry side. As Obama invited Luther to the stage, his anger translator offered a warning: "Hold on to your lily-white butts!" When Obama soberly intoned, "Despite our differences, we count on the press to shed light on the most important issues of the day," Luther's translation cut to the quick: "And we can count on Fox News to terrify old white people with some nonsense: 'Sharia law is coming to Cleveland! Run for the damn hills!' Y'all's ridiculous!"

Obama, through Luther, chided the press for its Ebola coverage and the Supreme Court for its support of excessive campaign financing. Luther also expressed Obama's anger about the failure of Republicans to grasp the challenge, and science, of climate change, even as Obama, as part of his routine, went uncharacteristically off the rails to channel his own anger: "I mean, look at what's happening right now. Every serious scientist says we need to act. The Pentagon says it's a national security risk. Miami floods on a sunny day and instead of doing anything about it, we've got elected officials throwing snowballs in the Senate" (a reference to Oklahoma senator James Inhofe carrying a snowball onto the Senate floor to prove that climate change is not real).

Luther attempted to calm Obama in black vernacular: "Okay, I think they got it, bro." But the president pressed on: "It is crazy! What about our kids? What kind of stupid, short-sighted, irresponsible bull—" Before Obama could finish the implied epithet, Luther interrupted. "Whoa, whoa, whoa, whoa!" he chimed in. "What?" Obama said. "Hey!" Luther barked. For Luther it was the last straw; the anger translator was outdone by the anger of the man he spoke for: "All due respect, sir, you don't need an anger translator. You need counseling. And I'm out of here, man. I ain't trying to get into all this." As he left the stage, Luther leaned over to audibly whisper to Michelle Obama, "He crazy." There was little doubt that Luther, and now Obama's own shtick, expressed precisely the kind of views the president voiced in private, out of the public spotlight.

If it was rare for Obama to find the space to speak frankly about race and politics, another figure in his administration famously found his racial register while possibly channeling the president's views.

Race Holder

Eric Holder was a favorite target of the far right from the moment he stepped onto the national stage as the nation's first black attorney general. Although he did not adopt the cadence and inflections of black street discourse, to many of his opponents Holder might as well have been a gangsta rapper spouting outrageous verbiage. In remarks delivered to the Justice Department to commemorate Black History Month in February 2009, two weeks after taking office, Holder offered one of the most courageous and honest speeches on race in America by a political figure in quite some time. Holder's remarks were predictably quoted out of context and thus easily misrepresented and misinterpreted. His comment that, despite considering themselves a melting pot, Americans are "essentially a nation of cowards" when it comes to race provoked ire and outrage.[8] Part of the backlash against Holder's words had to do with the national self-image he so brilliantly mocked with a provocative turn of phrase.

"As I said in that speech, people come from their race-protected cocoons, where they feel safe," Holder told me in an interview. "Given the history of race in this country—the slavery experience, the segregation experience, the Jim Crow experience, the violence that's associated with all of those experiences—it's a hard thing. We've got to ask what we can draw from those experiences, and how they affect present-day life. Some people say it's best not discussed," he acknowledged. "They say, 'Let's just deal with the things that we see in front of us in the twenty-first century and go from here.' And I think we do a great disservice by crafting solutions without being cognizant of the history that led us to the place where we are, both in terms of how that history has had an impact on African Americans and other people of color—but also the impact that it's had on white folks in this nation. It's crippled both groups, I think, in a variety of ways."

Holder's critics missed a vital link: by calling attention to America's racial cowardice, Holder, by implication, was praising the nation's ability to be courageous on other fronts—in creating one of the greatest democracies ever, for instance, and in taking the lead in aiding the world's poor, or in electing the nation's first black president. It is precisely because America has been so great at other tasks that we must point to its comparatively dismal record on race. When it comes to advances in scientific research, for example, we have been heroic. When it comes to racial dialogue and honesty, we have been timid and unimaginative; we have been cowards. That Holder's remarks caused such disgruntlement among the privileged chattering classes, and among some of the masses as well, points up the difficulty of the task at hand—getting Americans to talk openly and forthrightly about race—while underscoring just how fragile is the American racial ego.

Holder told me that he understood how white people could be hesitant to speak openly about race because they might be portrayed as bigots: "I think there's a fear, even among well-intentioned, good people, that if you speak about something that is heartfelt, but perhaps not politically correct, that you will be labeled a racist." Holder noted that black people "talk about race all the time," adding: "I don't know if white people talk about it all the time. I'm not there when they have those conversations. But I suspect they do talk about it more than they do when there are blacks and whites mixed, because it's a difficult thing.

"If you're white," he continued, "you don't want to be seen as some Neanderthal—even by the questions that you might ask. Even if you legitimately want to know, 'Well, I don't understand, why do you feel that way?' You then can be perceived, at best, as naïve, and at worst as insensitive-slash-racist. Blacks may feel tempted to ask, 'How could you not understand that?'" But, said Holder, "the reality is, they don't. They're white. They may say: 'I've never lived that. I don't know what it's like to be seen in that way. I don't understand the pain that you talk about. I don't understand the anger that you feel. So share that with me.'"

Holder pointed to the controversial article by *Washington Post* columnist Richard Cohen, penned in July 2013 in the aftermath of the George Zimmerman not-guilty verdict, arguing that—despite not liking what Zimmerman did and hating that Trayvon was dead—he could "understand why Zimmerman was suspicious and why he thought Martin was wearing a uniform we all recognize." That uniform is the hoodie, and what it symbolizes is "the reality of urban crime in the United States." Except, one might note, when it's worn by millions of Americans who never commit crimes, including, most famously, a figure like white New England Patriots coach Bill Belichick. Cohen complained that he was "tired of politicians and others" claiming that his views made him a racist. "Where is the politician who will own up to the painful complexity of the problem and acknowledge the widespread fear of crime committed by young black males?"⁹

Holder contends that Cohen's words have a place in a robust and frank conversation on race. "I think his views are what we have to be prepared to say is okay to be part of the dialogue. I don't agree with what he said. But I don't think he should be labeled a racist and dismissed." Holder argues that the expression of views like Cohen's should be encouraged in the interest of healthy dialogue and countered with fact, not emotion. "That's an attitude that's got to be confronted, that's got to be dealt with, that's got to be disproved if we want to make progress. You have to have an environment in which people feel empowered to say the kinds of things that he said, but that help promote the dialogue. If we are dismissive of him, we are ignoring a feeling that I suspect many people feel — and, interestingly enough, I suspect it's one that many black people feel."

Holder argues that Cohen's perception of black crime is part of a "new euphemism" springing up among conservatives about so-called "black-on-black crime." Holder maintains that it is misguided and deceptive to underscore crime among black folk "to the exclusion of everything else," as many conservatives do. "And, yes, there is a problem with the fact that ninety-four percent of all the black people who are murdered in this country are murdered by other black people. But about eighty-five percent of all the white people who are murdered in this country are murdered by white people. So we get white-on-white crime, black-on-black crime, I suppose. On the other hand, you could just look at it as a question of crime. And that's a part of the conversation. The focus on one, to the exclusion of the other, just does not make a great deal of sense, and is not the way progress is ultimately made."

Holder's broader point is that black history is American history, and that we do further injustice to the unduly neglected contributions of our black citizens to the common good if we continue to revel in historical ignorance, or avoid addressing the racial trauma that makes the study of black history necessary. Holder understands that the heroic efforts of black activists and intellectuals in the past forced America to clarify its ideals in the crucible of struggle that helped to define the American character. From slavery to the civil rights movement, black social action offered the nation the opportunity—and template—to transform democratic rhetoric into uplifting deeds that brought America closer to its founding principles. It is that history that shaped Holder personally, and intellectually, too, and formed his beliefs about how he should conduct himself as the nation's first black attorney general.

"We're all products of our upbringing and the experiences that we've had," Holder said to me. "I was a black kid raised in New York City during the civil rights movement, although I didn't get to participate in it. I was a little young. But I certainly observed it. And being married to the sister [Sharon Malone] of one of the great warriors in that struggle, Vivian Malone [one of the first two black students in 1963 to enroll in the University of Alabama, where Governor George Wallace infamously blocked her path], my sensibility is going to be different than a white person raised in California who did not have to deal with what it means to be a black man. And it doesn't mean that I had to be discriminated against myself. Doesn't mean that I had to have my home firebombed. When I was a kid and I watched those kids in Birmingham being hosed and being attacked by dogs, that had an impact on me in a way that it would not have had, I think, on my white counterparts in New York City watching the same footage." Holder said that he brought "that sensibility to the job" of attorney general. It was "this notion of social justice, this

notion that looking at things racial from a different perspective matters. And it's natural, given who I am, and the experiences that I've
had, and given the way African Americans have had to deal in this
country for the last three, four hundred years."

Holder in his Black History Month speech also argued that other
movements for social change — the feminist and antiwar movements
most notably — took their cue from the noble efforts of often unsung
black Americans who gave their lives unselfishly to improve the lot
of their brothers and sisters, and thus the lot of the nation. Holder
was not interested in a one-way conversation; he admitted that there
is legitimate space in the culture to debate the nuances and complexities of affirmative action. But without telling the truth about the
bloody history that made affirmative action necessary, we can hardly
debate its finer points of application or abuse. In making such arguments, Holder insisted that black history is not only good for black
Americans but also vital for the entire country. Unlike others who
lament black history's segregation into a month, and therefore seek
to get rid of the celebration, Holder saw the need for weaving black
history into the fabric of ordinary American life and offering our
citizens a critical tool of self-inventory, while broadening the horizon
of our nation's self-awareness around racial issues.

Holder understands that America's segregation of its history reflects its history — and its present practice — of segregating its people. The attorney general offered an alert and truthful account of
how American social interaction follows patterns established some
fifty years ago, when the nation suffered the effects of Jim Crow,
with its mandate for legal separation of the races. Now that official
restraints have been loosened, Americans nonetheless retreat into
"race-protected cocoons" and rely on clichés, stereotypes, and familiar habits of ethnic and racial congregation, instead of opening
ourselves to the uncomfortable but rewarding prospect of facing
one another across the chasm of learned behaviors and inherited
reflexes.

Race Man and a Race's Man

Holder's fair and tough-minded speech on race called attention to the comparatively tame race rhetoric of his black boss in the Oval Office. Holder's reputation as an attorney general who carried out his duties while tending to his racial roots cast him in a more favorable light, in some black quarters, than President Obama. Holder fought vigilantly to protect the Voting Rights Act even after the Supreme Court in 2013 had profoundly weakened it in *Shelby County v. Holder*. "We moved all the resources that we had in Section Five cases [the part of the Voting Rights Act that explains how the requirement for federal approval of changes to election laws—called preclearance—works] to Section Two [which prohibits discrimination in voting practices or procedures] of the Voting Rights Act," Holder told me, "and worked with members of Congress to, in essence, reverse the Supreme Court's very flawed decision and put together coalitions to give life again to Section Four." (Section Four established a formula to identify which areas of the country—nearly all in the South—were most likely to experience racial discrimination in voting; this section was ruled unconstitutional by the Supreme Court in *Shelby County v. Holder*.)[10]

Holder fought to reduce the racial disparity in sentencing for nonviolent drug offenses as well, a chasm opened up by a former boss, Bill Clinton, in his hugely detrimental 1994 crime bill. Holder also sued states like Texas and North Carolina over discrimination against blacks in their voter ID laws. And he visited black communities, like Ferguson, Missouri, that faced the consequences of poor policing, while forcing its police department, like many more throughout the nation, to address its racist practices. Cultural critic Rich Benjamin colorfully dubbed Holder Obama's "Dark Doppelganger," his "Inner Nigger"—arguing that Holder could say and do black things that Obama could scarcely imagine the freedom to say and do.[11]

"I'm not his Inner Nigger," Holder insisted to me. "We're two

black men who happen to share a worldview." Holder made a surprising observation about what things might have looked like if
he and Obama had switched political places. "If he were attorney
general of the United States, he would be an attorney general in
my mold. And I think if I were president—and that's a frightening
thought—I'd probably be a president like him." That seems to fly
in the face of the evidence that Holder is far more conscientious
about drinking from the roots of his black identity than the much
more reluctant Obama. But Holder insists that is not true, defending
Obama's sparse race rhetoric since becoming president.

"I'd husband my racial capital," Holder says. "Spend too much and
you devalue it. Spend it not sparingly, but wisely, and you increase
its worth." Holder defends Obama's approach, saying that he as attorney general had greater latitude to express himself, while Obama
faced different constraints. "I'm not the president of the United
States. Every word of every conversation that he has is looked at,
dissected, criticized, praised. And so he's got to be careful. He cannot be seen as a polarizing figure. That would not serve him or the
country well."

What of the need for presidents to speak courageously about race?
"I think there is a responsibility of any president—not just him—to
speak truth. And I don't think that the criticism that he's been reluctant to do that is justified." Holder's argument might be difficult to
defend if we examine the frequency or efficacy of Obama's racial
discourse. Still, Holder says that it was Obama who supported him
while he was the nation's chief lawyer.

"When I asked for all of these resources to do the civil rights enforcement, some people gave me a great deal of credit. Well, guess
what? He's got to make all these decisions about how he's going to
use the resources that he has. 'Am I going to put them in HUD, am
I going to put them in Defense, Education?' He gave me in that first
year the largest increase in the history of the Civil Rights Division,
in terms of lawyers and money," says Holder, "which enabled us to
do the things we've done on voting rights and police misconduct

cases. We filed record numbers of cases as a direct result of him saying, 'I'm going to make that a priority.' He doesn't run the Justice
Department. But he enabled me to do what I did. I think people are
a little naïve to think that Eric Holder was out there operating as a
free agent, not tied to Barack Obama."

Obama's relentless chiding of black America didn't bother Holder
either, though he argued that balance is the key to making such
speeches effective. "I think that those speeches where you take black
people to task are appropriate. And the notion of personal responsibility is something that I've talked about — as long as you couple that
with the notion of societal responsibility. Everybody has got to play
a role, but the nation has a role to play as well. There are institutional
barriers that have to be taken down. There are attitudinal things that
have to be changed, coupled with the notion of individual responsibility." Ultimately, he says, "I think that you have to make it an
inclusive conversation, where you're giving tough love to the African
American audience but at the same time you're doing the same thing
to the society in which we all live."

While others have pointed to the glaring absence of such balance in Obama's speeches, Holder argued that the president's nod
to structural forces was not nearly as newsworthy or sensational. "I
think the balance is there, but that's not a sexy thing. That part of
the speech doesn't get reported on; [it] doesn't get commented on.
The thing where the black president says tough things to the black
community, that's the headline. It was the same with my talk; it's
known as the 'nation of cowards' speech. That was the fourth, fifth,
sixth line in the speech. But it's far more nuanced than that. It's far
more optimistic. But it's not sexy. It's easy to say, 'Attorney General
says America is a nation of cowards.'"

The big difference, of course, is that Holder was emphasizing a
societal deficit in calling for the collective responsibility of the country, and he was, at least in part, holding white people accountable
for lacking the courage to confront race, strikingly dissimilar in tone
and approach to the president. Obama gently rebuked Holder for

his comments, yet the former attorney general insists that Obama shifted to holding white people accountable in his Trayvon Martin speech.

"When he talked about 'White America, you need to understand where we are coming from,'" Obama was, Holder says, making a demand for the broader culture to engage black America and understand its plight. "That was the thing that was different about that speech. He was not speaking *about* the black community; he was the spokesman *for* the black community. Which was, I think, a little different." This time Obama "was taking white America to task, to say, 'Look, this is why we see the Zimmerman verdict the way that we do. And you've got to understand that. This is where we are coming from. The lens that we use in viewing the world is different than the lens that you have. Not right, or wrong—it's different. You need to understand the pain that we feel. What it's like to be a man who is humiliated, made to feel like less than a man when you are viewed as a criminal. When you're stopped for no reason."

It was indeed a stretch for Obama to defend rather than assault black folk, but in his Trayvon Martin speech, undeniably poignant, he did not take white America to task in the same fashion in which he has lambasted black communities. Holder is satisfied that his personal knowledge of the president offers a far more nuanced and complicated portrait of a man who takes pride in his identity.

"People don't know the president the way I know the president," Holder protests. "He's a person comfortable in his skin and conscious of the history of this nation, and determined to change it. He's a proud black man. He doesn't shrink from that. He sees that he has a special responsibility, being the first. I felt that on a much smaller scale." As president, Obama "feels a special responsibility to make sure that he represents our people in the best tradition of firsts, like Jackie Robinson. People tend to forget: Jackie Robinson, in the early part of his career, difficult as it was, just took it. Probably resulted in his early death—all the stuff that he had to put up with. And I know that the president takes a lot of stuff in a political way. When you question basic things about him, like where was he born, with the birthers. He just takes it. And that, I think, is a sign of strength."

But, Holder stresses, Barack Obama is "a person who's conversant with the history of black people in this country, the history of Africans, the colonial experience. He's not a person who is uncomfortable with his blackness. He understands who he is. He understands the symbolic importance of what he's accomplished," says Holder. "And I think he wants to be successful on his own, but also lay a foundation so that others might follow. Through his example he will make things better for those who will follow. And I don't mean follow as president, but who follow him as a young black male. He wants to make the world a better place for black kids like my teenage son."

Domestic Terror in Obama's America

Making things better for black teens is far more urgent because of the charged climate for black folk in the second decade of the twenty-first century, when it is clearer than ever that black death is, once again, as it was for much of the twentieth century, both sport and lust—whether at the hands of the police or ordinary citizens like Michael Dunn, who killed unarmed Jordan Davis in Jacksonville, Florida, or Dylann Roof, who massacred nine blacks in a church in Charleston, South Carolina. That these actions are occurring now underscores the paradox—and in some ways the predictability—of the resurgence of white terror during our first black presidency.

Outraged motorist Dunn killed seventeen-year-old high school student Davis in 2012 at a gas station after a verbal altercation over loud music coming from the car in which the teen was sitting with three other black teenage males. There is a hoary tradition of violence in the nation to snuff out even the tiniest gesture of black defiance, whether a raised voice, an arched eyebrow, or an educated tongue. From this grew the Olympics of annihilation that prompted willing whites to compete for the honor of crushing black life. The bullets that racked Jordan's body were set in motion long before they spun out of the barrel of Dunn's gun in Florida. The reasons are as irrelevant as they are random for this murder, and for so many others like it: obscene couplets of rap music, suspicious clothing, and a threatening demeanor—troubling symptoms, it seems, of simply being black, an offense punishable by death.

The loud music that shook the Dodge Durango in which Jordan was seated was no match for the dark racial sentiment that rattled the heart of his killer. Dunn's prejudice stained the letters he sent from prison as he awaited trial, echoing a shopworn gripe that "gangster-rap, ghetto talking thug 'culture' that certain segments of society flock to is intolerable." (Beatty found it irresistible; Dunn found it unforgivable.) Dunn let fly a bigoted tirade against his black fellow inmates, who "all act like thugs," saying hatefully that "if more people would arm themselves and kill these [expletive] idiots when they're threatening you, eventually they may take the hint and change their behavior."[12]

It is sad, even tragic, that so many blacks believe they can get rid of the "thug" label by denouncing rap music, sagging pants, and the boorish behavior of a few wayward youth. (That is another reason why it is unfortunate when politicians like President Obama readily resort to the term "thug" to brand all participants in urban uprisings; it plays too easily into a history of demonizing black behavior and identity that reaches far beyond social resistance to black inequality on the streets of America.) The cold fact is that all blacks are in the same boat no matter their pedigree or performance. Nothing except self-erasure will satisfy those who despise blacks.

Racists like Dunn have little interest in getting the facts right. They scarcely make the effort to tell one black person from the next. Seattle Seahawks cornerback Richard Sherman was seen as a thug because he was brash and dramatically self-assertive in a postgame interview after the NFC Championship contest in 2014. His Stanford University degree did not spare him racist rants in social media. Even-tempered and fair-minded President Barack Obama has been called a "Chicago thug" by right-wing radio provocateur Rush Limbaugh. Professional golfer and former Ryder Cup captain Paul Azinger called basketball superstar LeBron James a thug for a bit of trash-talking (though the king of trash talk, Larry Bird, was never viewed as a thug during his playing days). And black teen Trayvon Martin was widely mislabeled a thug after his brutal killing by George Zimmerman, for whom a far more compelling case of being a thug could be made, particularly given the violent nature of his post-verdict behavior.

Dunn called Jordan a thug without a shred of evidence, except, of course, for his blackness itself, a blackness Dunn deemed too loud and defiant, a blackness with the backtalk built in. In judging Jordan to be a thug, Dunn became a thug himself. But he forgave himself in advance by convincing himself of his righteousness and racial innocence. This is the infuriating irrationality brought on by unchecked racial aggression. Members of a dominant group often portray themselves as victims. They excuse their ethical lapses and appeal to their fears and anger to justify their hateful actions. They often project their moral flaws onto their victims and then target them for social or literal death.

Dylann Roof was such a figure: the murderous rage of the lone-wolf white supremacist led him to unload his .45 pistol on June 17, 2015, on nine black souls in Charleston at the Emanuel African Methodist Episcopal Church, seeking, he confessed, to provoke a race war. Roof's act of race hate is a tragic monument in the landscape of white racial terror. Sites and spaces of black life have come under attack from racist forces since the nation's founding, but the black church is a unique target. That is because of its enormous symbolic value: the black church is not just where black people gather but a space where the sacredness of black existence finds vocal affirmation. In too many other places, black self-worth is bludgeoned by bigotry or hijacked by self-hatred: the notion that black culture is too dumb, or black lives too worthless, to warrant the effort to combat black people's enemies. The black sanctuary breathes in black humanity while the pulpit exhales unapologetic black love. For decades these sites of love have been magnets for hate and white terror.

Near the beginning of the engrossing 2014 film *Selma* there is an especially harrowing scene: four black girls, adorned in their 1963 Sunday best, and cloaked in the sweet innocence of youth, descend the staircase of the Sixteenth Street Baptist Church in Birmingham, Alabama, before they are bombed to their heavenly reward far before their time. In that same city, in 1958, a dynamite bomb rocked the Bethel Baptist Church, led by the Reverend Fred L. Shuttlesworth, a civil rights luminary. It would take more than two decades to bring the white supremacist perpetrator to justice. As the drive to register black voters heated up during Freedom Summer in 1964, nearly three dozen black churches in Mississippi were bombed or burned.

The hatred of black sacred space didn't end in the sixties. In July 1993, the FBI uncovered a plot to bomb the First AME Church in Los Angeles, wipe out its congregation with machine guns, and then assassinate Rodney King—the black motorist whose videotaped beating by four Los Angeles cops led to charges of assault with weapons and excessive force, of which the police were acquitted, sparking an urban rebellion in L.A.—in hopes of provoking a race war, much like Roof. In 1995, several men took sledgehammers to the pews and kitchens of black churches in Sumter County, Alabama. A year later, the Inner City Church in Knoxville, Tennessee, was bombarded with as many as eighteen Molotov cocktails as its back door was splashed with racist epithets. President Clinton appointed a task force in 1996 to investigate church fires, which by 1998 had singed the holy legacies of 225 black houses of worship. In November 2008, three white men set the Macedonia Church of God in Christ in Springfield, Massachusetts, ablaze hours after Barack Obama was elected the nation's first black president. And six black churches across the South were burned down in June 2015.

Roof's terrorist massacre wasn't the first time Emanuel AME

Church faced racist violence. In 1818, black religious leader Morris Brown left a racially segregated church in Charleston and formed the African Methodist Episcopal Church of Charleston, which later became Emanuel AME Church. After Denmark Vesey, one of the church's founding members, plotted a slave rebellion, and was foiled in the effort by a slave who betrayed his plans, Emanuel was burned to the ground by an angry white mob.

Despite this history, black churches are generally open and affirming of whoever seeks to join their ranks — unlike white churches, which have often been rigidly divided along racial lines. The AME Church was born when founder Richard Allen spurned segregation in the white Methodist Church and sought to worship God free of crippling prejudice. Early church leaders took seriously the scripture of Acts 17:26, which claims of God, "From one man he made all the nations, that they should inhabit the whole earth," even as they embraced the admonition in Hebrews 13:2, "Do not forget to show hospitality to strangers, for by so doing some people have shown hospitality to angels without knowing it."

That is how it is possible that the doors of Emanuel African Methodist Episcopal Church were open to Roof, a young white participant, who, after an hour of prayer, raised a weapon and took nine lives. One of the survivors told Sylvia Johnson, a cousin of murdered pastor — and state senator — Reverend Clementa Pinckney, that Roof argued: "I have to do it. You rape our women and you're taking over our country, and you have to go." The vortex of racist mythology spun into a plan of racial carnage and terror.

Roof's maniacally worded manifesto articulated his rancid racist worldview. He admits that he wasn't reared in a racist environment, but that he became racially aware after the Trayvon Martin case. "I kept hearing and seeing his name, and eventually I decided to look him up," Roof writes, reading "the Wikipedia article and right away I was unable to understand what the big deal was. It was obvious that Zimmerman was in the right." Roof says he was prompted to "type in the words 'black on White crime' into Google, and I have never been the same since that day." He was directed to the website of the Council of Conservative Citizens, where he found "pages upon pages of these brutal black on White murders. I was in disbelief. At this moment I realized something was very wrong."[13] Just as with perpetrators of terror in the political world, Roof was radicalized by his exposure to racist websites and their hateful pedagogy. They exacerbated Roof's burgeoning antiblackness that led to the violent racist terror when he stepped into the Emanuel AME Church.

The black church is a breeding ground for leaders and movements to quell the siege of white racist terror. From the start, black churches sought to amplify black grievance against racial injustice and to forge bonds with believers to resist oppression from the broader society. The church's spiritual and political missions were always intertwined: to win the freedom of its people so that they could prove their devotion to God. Some critics see the black church as curator of moral quiet in the face of withering assault. Religious people are accused of being passive in the presence of social injustice—of seeking heavenly reward rather than earthly justice.

In truth, the church at its best has nurtured theological and political resistance to the doctrine of white supremacy and the forces of antiblack hatred. The church has supplied leaders and blueprints for emancipation—whether in the preaching of Frederick Douglass or Prathia Hall or in the activism of Martin Luther King Jr. or Al Sharpton. But the church is also the place where black people are most vulnerable. Oddly, stereotypes of the sort that Roof nursed are unmasked in such a setting. It is not murderous venom that courses through black veins but the loving tolerance for the stranger that is the central moral imperative of the gospel.

Adherence to the moral imperative to treat strangers kindly may have led to the black parishioners' death in Charleston. Roof thus exploited the very kindness and humanity he found before him. The black folk gathered in that church were the proof that he was wrong;

they were the living, breathing antithesis of the bigoted creeds cooked up in the racist fog he lived in. It was not their barbarity but the moral beauty of black people that let an angel of death shelter in their religious womb. Its openness and magnanimity are what make the black church vital in the quest for black self-regard.

When preachers like Clementa Pinckney stand in the house of God to deliver the Word, they embrace the redemption of black belief—a belief in self and community. In a country where black death is normal, even fiendishly familiar, black love is an unavoidably political gesture. And that is what happens in black churches. The unavoidably political act of black love—an act that seems to make black houses of worship a target of hate. It is a political act in this culture that must remind the nation, once again, as hate and terror level black communities, that black lives matter.

Déjà Vu All Over Again

The Charleston massacre forced Obama to speak again, wearily, on race, though, as in many instances before, the situation seemed to strain him; his initial response was noteworthy for the sense of resignation glued to his face and weighing on his voice. Speaking in the White House on June 18, 2015, from the James S. Brady Press Briefing Room—an irony that should not be lost on us, since Brady too was shot, and permanently disabled, in the assassination attempt on Ronald Reagan in 1981, and became a vigorous advocate for gun control—Obama made a bigger deal of gun violence than of race in his comments.

"I've had to make statements like this too many times," the president said.[14] "Communities like this have had to endure tragedies like this too many times . . . because someone who wanted to inflict harm had no trouble getting their hands on a gun." Obama continued, "This type of mass violence does not happen in other advanced countries . . . with this kind of frequency." Obama said that the fact that the massacre happened in a black church "obviously raises questions about a dark part of our history." He noted that this was "not the first time that black churches have been attacked," but then quickly struck a universal chord in saying that "hatred across races and faiths poses a particular threat to our democracy and our ideals."

Near the end of his statement, the president let Dr. King speak for him as he quoted the civil rights leader's words of eulogy for the four little girls killed in the Birmingham church bombing: "They say to us that we must be concerned not merely with who murdered them, but about the system, the way of life, the philosophy which produced the murderers." King nodded to a systematic oppression that Obama often has difficulty identifying and opposing.

The next day, before a meeting in San Francisco of the U.S. Conference of Mayors, Obama disputed the notion that he had appeared resigned to having no successful legislation for curbing gun violence, declaring, "I'm not resigned. I have faith we will eventually do the right thing," because one "doesn't see murder on this kind of scale, this kind of frequency in any other advanced country." The president raised again the Newtown tragedy, which he has said was the worst moment of his presidency. Obama briefly acknowledged that "racism remains a blight we have to combat together," and while the nation "has made great progress," we "need to be vigilant, because it lingers . . . and it betrays our ideals and tears our democracy apart."[15]

Later that day Obama got racially unshackled. He may have been freed by the far more direct and extensive comments of 2016 presidential candidate Hillary Clinton. The former secretary of state took to the podium after Obama and offered the mayors, and the country, a far more comprehensive engagement with the racial politics engulfing Charleston and the nation. Clinton anchored her comments in black history; she acknowledged that African Americans had celebrated, the day before her speech, Juneteenth, "a day of liberation and deliverance" when, as "President Lincoln's Emancipation Proclamation spread from town to town across the South, free men and women lifted their voices in song and prayer."[16] Clinton expressed confidence that the black folk in Charleston would draw on their faith and history to see them through: "Just as earlier generations threw off the chains of slavery and then segregation and Jim Crow, this generation will not be shackled by fear and hate."

Clinton addressed gun control too, but refused to separate the politics that prevent sensible and effective legislation from the swirling racial currents that flooded Charleston and the nation in the ugly sweep of violent animus—what Clinton later called "racist terror." Clinton said in her speech that, tragically, despite our devotion to human rights and diversity, "bodies are once again being carried out of a black church," and that, once again, "racist rhetoric has metastasized into racist violence." Clinton argued that "it is tempting to dismiss a tragedy like this as an isolated incident, to believe that in today's America, bigotry is largely behind us, that institutionalized racism no longer exists. But despite our best efforts and our highest hopes, America's long struggle with race is far from finished." She acknowledged that race is a difficult issue to talk about and that "so many of us hoped by electing our first black president, we had turned the page on this chapter in our history," and that "there are truths we don't like to say out loud or discuss with our children," but that "we have to" because it is "the only way we can possibly move forward together."

Clinton laid out the facts: blacks are nearly three times more likely than whites to be denied a mortgage; the median income of black families is $11,000, while for whites it is $134,000; nearly half of black families have lived in poor neighborhoods for two generations, compared to just 7 percent for whites; black men are more likely to be stopped and searched by police, charged with crimes, and sentenced to longer prison terms than white men, 10 percent longer than white men for federal crimes; black students suffer from the vast re-segregation of American schools; and black children are 500 percent more likely to die from asthma than white children. Clinton lamented:

More than half a century after Dr. King marched and Rosa Parks sat and John Lewis bled, after the Civil Rights Act and the Voting Rights Act and so much else, how can any of these things be true? But they are. And our problem is not all kooks and Klansmen. It's also in the cruel joke that goes unchallenged. It's in the off-hand comments about not wanting "those people" in the neighborhood. Let's be honest: For a lot of well-meaning, open-minded white people, the sight of a young black man in a hoodie still evokes a twinge of fear. And news reports about poverty and crime and discrimination evoke sympathy, even empathy, but too rarely do they spur us to action or prompt us to question our own assumptions and privilege. We can't hide from any of these hard truths about race and justice in America. We have to name them and own them and change them.

Clinton's remarkable oration was steeped in black culture and charged with sophisticated analysis, and was a remarkably honest reckoning, by a major American politician, with both intimate and institutional racism—racism of the heart, and racism in the systems of society.

Clinton's speech was only one indication of the rapidly changing political atmosphere in the wake of Charleston. White politicians such as South Carolina's Republican governor Nikki Haley and Republican senator Lindsey Graham called for the Confederate flag—the flag that Dylann Roof embraced as a symbol of his unquenchable hate—to be removed from the State House grounds. White southern politicians in Mississippi, Tennessee, Maryland, Virginia, and North Carolina also pledged to remove their Confederate flags, or to ban their depiction on license plates, part of an "emotional, nationwide movement to strip symbols of the Confederacy from public parks and buildings, license plates, Internet shopping sites and retail stores."[17] It was a stunning turn of events in southern culture and politics, and the death of nine black citizens was the reason for the change.

Don't Call It a Comeback

The colossal sorrow that engulfed the nation when nine souls perished in a southern church may have emboldened Obama to engage race more honestly, and perhaps in a slightly more unguarded fashion. On the Friday of Clinton's speech, as the nation endured a swift racial transition into an explicit confrontation with the symbols of hate in the form of the Confederate flag, Obama recorded an hourlong interview for comedian Marc Maron's podcast series *WTF*. The podcast is usually taped in Maron's Los Angeles home garage, nicknamed "the Cat Ranch." Obama came to Maron's garage for an interview that aired the following Monday and garnered a great deal of notice, primarily because of what he said about race.

"I always tell young people in particular, 'Do not say that nothing's changed when it comes to race in America unless you lived through being a black man in the 1950s, or '60s, or '70s,' " Obama told Maron.[18] "It is incontrovertible that race relations have improved significantly during my lifetime and yours. And that opportunities have opened up, and that attitudes have changed. That is a fact. What is also true is that the legacy of slavery, Jim Crow, discrimination in almost every institution of our lives—that casts a long shadow. And that's still part of our DNA that's passed on."

That is nothing Obama hadn't said in a few of his commentaries on race. But what followed was something he had never said in public since becoming president, and it created a big controversy. "We are not cured of [racism]. And it's not just a matter of it not being polite to say *nigger* in public. That's not the measure of whether racism still exists or not. It's not just a matter of overt discrimination." Obama's use of the racial epithet was brave; he meant to draw attention to the vicious legacies of racism that persist even when that epithet is not used. By using the real word instead of its abbreviated euphemism, "the N-word," Obama conjured a visceral sense of history: he used the ugly term white supremacists often used when they lynched, castrated, or drowned black people. Dylann Roof used the word in his white terrorist manifesto. Obama offered a one-word lesson in the bitter history of race encapsulated in a single utterance that is widely considered, in América at least, the vilest term in the English language.

Obama also underscored a racial lesson the nation routinely overlooks: the absence of the word "nigger" does not guarantee the presence of racial justice; refraining from pronouncing the word in public does not purge racist practices. Neither does the absence of overt discrimination suggest that the nation is free of racism. By risking the controversy he certainly knew that his use of the word would—and did—provoke, Obama appeared to welcome a racial bluntness that his advocates have hungered for, and that the nation has desperately needed. Some deemed Obama a perpetrator of racial insensitivity rather than racism's critic. That misses a crucial point: many presidents have uttered the term, though rarely in public, and none with scare quotes around the word as Obama intended; they have mostly used it as white racists have deployed the term over the years, either as an explicitly derogatory term or, equally problematic, as a term of racist neutrality, a term meant to describe all black people.

This appeared to be the beginning of a *Bulworth* moment when Obama might tell the truth and open the avenues of forthright racial conversation—when he might draw on Beatty, Holder, and even Hillary Clinton, or his wife, Michelle, in forging his analysis. It seemed that Obama was finally willing to forgo the politics of respectability that strictly forbade the utterance of the term "nigger," even by a man who knew the revulsion the term evokes, and its taboos, and its conflicted use by blacks themselves, and who only said it a single time in public to emphasize his point. Yet as soon as he uttered the inflammatory and provocative word, he fell back into a far safer pattern of racial exposition in his *WTF* interview, proving that he was not yet ready to take the leap he secretly craved. Obama repeated familiar themes in the interview: that cops have a tough job and we can't expect them to solve our racial problems; and that early education can break the poverty cycle. This is hardly a bold-speaking politician willing to tell truths that black people are often forced to learn but from which white Americans can easily escape. If this wasn't Obama's *Bulworth* moment, something far more promising was soon to come.

Free at Last?

When President Obama took the podium in July 2015 at the annual convention of the NAACP in Philadelphia, it sounded as if Michelle Alexander, author of *The New Jim Crow*, and Eric Holder had hacked his computer and collaborated on his speech.[19] It may not have been a full-on *Bulworth* moment, but it was decidedly closer than the one-off racial epithet of his *WTF* interview. "By just about every measure, the life chances for black and Hispanic youth still lag far behind those of their white peers," Obama said in his address on criminal justice reform.[20] He pointed to "a legacy of hundreds of years of slavery and segregation, and structural inequalities that compounded over generations. It did not happen by accident." After underscoring the ill effects of bigotry in employment, schools, and housing, Obama argued that the criminal justice system "remains particularly skewed by race and by wealth." Blacks and Latinos "make up 30 percent of our population; they make up 60 percent of our inmates." Obama spoke of how one in thirty-five black men, and one of every eighty-eight Latino men, is serving time, while the number is only one in 214 for white men. "The bottom line is that in too many places, black boys and black men, Latino boys and Latino men experience being treated differently under the law."

By this point President Obama had not simply changed his tune of avoiding race; he seemed to be singing in a far more comfortable register. Not only was he speaking out on the racial disparities that cripple the justice system. He also helped to clarify the definition of rape when asked a question about beleaguered entertainer Bill Cosby at a July 2015 press conference, thereby implicitly criticizing the comedian, who allegedly administered drugs to more than thirty women and had sex with them over a forty-year period.[21] Obama also sought to aggressively enforce legal bans on residential discrimination, making cities and localities accountable for the use of federal housing funds to reduce racial disparities, or else face penalties if they failed to comply. This was part of Obama's effort to ensure that the civil rights–era dream of fair housing came to pass.

The president also finally became more willing to grant pardons to prisoners who were often unjustly saddled with life sentences for nonviolent drug offenses. Obama dramatized their plight by becoming the first sitting president to visit a federal prison. He restored diplomatic ties with Cuba, a historic achievement that thawed relations between the two countries after fifty-four years of antagonism. Obama also stood in the storied black Ninth Ward of New Orleans on the tenth anniversary of Katrina, and instead of arguing, as he had a decade earlier, that the government's criminally slow response to the disaster was a matter of color-blind ineptitude, Obama acknowledged in an anniversary speech that New Orleans "had for too long been plagued by structural inequalities that left too many people, especially poor people of color, without good jobs or affordable health care or decent housing."[22] He restored the Alaskan native name for North America's tallest peak when he changed it from Mount McKinley to Denali, thereby, according to one scholar, engaging in "an act of decolonization."[23] And he even partially redeemed his first trip to Africa as president with a 2015 visit to Ethiopia and his paternal homeland of Kenya—where Obama publicly performed an African dance, stood up for LGBTQ Africans, criticized genital mutilation, encouraged African leaders to adopt term limits, embraced his African kin, and proudly proclaimed, "I'm the first Kenyan-American to be president of the United States," a statement that would have been unthinkable before his 2012 reelection, and was a not-so-subtle needling of the birthers who viciously trafficked in conspiracy theories about his Kenyan birth.[24]

Obama's racial renaissance was sparked, in large part, by the vagaries of history and the play of contingent elements that reveal, even force, a president's hand, rushing him to the bully pulpit in ways that only a few months before might have been inconceivable. The siege of black death across the land was key, too—police killings of unarmed black citizens, the vicious lash of white terror at churches and gas stations. The rise of the Black Lives Matter movement in the streets also pushed the president in the right direction, along with a steady undercurrent of principled black criticism. Of course, the toxic racial atmosphere when Obama first came into office put the first black president in an untenable position: if he paid too much attention to the racist attacks, he looked weak and ill-prepared for the predictable racial blowback.

"I am subject to constant criticism all day long," the president told me in our Oval Office discussion. "And some of it may be legitimate; much of it may be illegitimate. Some of it may be sincere; some of it may be entirely politically motivated. If I spent all my time thinking about it, I'd be paralyzed. And frankly, the voters would justifiably say, 'I need somebody who's focused on giving me a job, not whether his feelings are hurt.'" Obama's racial stoicism does not mean that he hasn't borne the brunt of racist feelings that spread throughout the land, especially those that unfurled under the banner of the Tea Party. Obama acknowledged to me that a great deal of the resistance he faced from the Tea Party had more to do with antigovernment emotions than with strictly racial animus, even though he understood how the two intertwined. "Are there probably elements within that movement that focus on my race? I think that's probably the case. I don't remember any other president who was challenged about where he was born despite having a birth certificate."

If the political resistance to Obama has often been racially driven, plausible deniability has also run high as right-wing politicians claimed that ideology more than race motivated their bitter fulminations against Obama's policies. This often felt to black people like a thin disguise for old-school racial demonology. The remnants of barely suppressed racial hostility cluttered the political and cultural landscape for many black witnesses of Obama's racially captive presidency. It was a painful paradox: the most powerful man in the world hindered by a thousand cords of racial resentment strung by elected officials and right-wing media. He was a black Gulliver to their Lilliputian whiteness. Obama therefore often practiced the politics of racial sublimation: he took the energy of race and redistributed it over the political landscape in a kind of racial détente that forged an uneasy alliance of racial amnesia and racial avoidance. Still, no matter the calm on the surface, racial tensions percolated beneath. When they finally erupted in police killings and racial terror and black resistance, Obama's path to public proclamation was cleared.

Bracing racial rhetoric, in tandem with targeted public policy, can make a big difference in how race is lived, a lesson Obama seems to have learned afresh in the last quarter of his presidency. The president's push for prison reform was one example, as he argued that instead of devoting $80 billion to incarceration, we should invest in pre-K education and jobs for teenagers, both of which would repay the investment far more grandly than a life diminished behind bars. He also argued, as others have done, including Children's Defense Fund founder Marian Wright Edelman, that we should interrupt the school-to-prison pipeline. Obama argued for vast improvement in prison conditions. It seemed that his urge for reform could be aided by a complementary push to restore educational alternatives for prisoners and to beef up training programs that integrate ex-offenders into society. In light of the hostilities between communities of color and the police, Obama summoned a group to make recommendations, some of which he adopted, such as the demand for police body cameras.

Each effort was commendable, and in some ways overdue. But something bigger is called for: we need a new Kerner Commission report that is updated for our day, paying special attention to how black folk are viciously targeted by unethical police practices. It is true that calling for a commission might not seem like the most systematic fix, but a serious investment in assessing the state of inequality and ingrained racism in America will show us clearly what work remains. It will be harder to ignore, less ephemeral than mourning or protests.

In our conversation in the Oval Office, the president told me that he aimed to speak about universal values. "If I'm going to prophesy, I want it to be about values that everybody can rally around," Obama said. "And I think that's always been the strength of the African American movement. At its base, what's always been strongest about the civil rights movement has been when it said, yes, there is a unique problem here that arises out of race and slavery and segregation. But when you lock us up, you're imprisoning yourself in some fashion. When you deprive that child of opportunity, you're weakening yourself."

Obama's revived racial rhetoric felt like a *Bulworth* moment—in fact, plenty of them—and a balm to those who yearned for his presidential leadership. In November 2014 the nation witnessed the spectacle of a black president holding a news conference on one side of a television split screen while in Ferguson, tear gas and sirens swirled around a crowd protesting the failure of a grand jury to indict a police officer in the death of an unarmed black teenager. Obama was stern, disciplined, his handsome face strained by the remorseless thrum of events—events that possessed a punishing repetitiveness, a relentless logic of black suffering as their end. The president appeared to be, as he had often stated, *not* the president of black America, but instead the president of the United States of America, now disuniting as he spoke, his words metaphorically met by ringlets of smoke escaping the fires that raged in place of understanding and hope.

In the year following Ferguson, a seismic event that fractured the nation's racial landscape, Obama found a way to be the president of all America while also speaking with special urgency for black Americans. That was a crucial development. Obama's disclaimer of not being the president of black America meant that a significant segment of the population didn't feel his state-sanctioned concern. Of course it is true that it is not just in the official exercise of power that representation matters; it matters as well, for black people at least, in the signifying gestures of support that a president can offer. It is undeniable that presidential attention to a population and its issues can buttress the belief that democracy is for all Americans. It is, for black people, part of the folklore of presidential favor from which they have often been excluded; such favor is delivered in the gritty interstices of the machinery of politics, in the grooves and treads where race has worn down democracy, and where the grease of small accommodation, or minor concession, oils the operation of trust and investment, particularly for those against whom the machine has been historically poised.

"That moral force is not just a matter of who's getting what," the president said to me in our Oval Office conversation, "but it's rather: 'Is everybody, as part of the American family, sitting at the table?' And that naturally means that, as we now sit at the table, we don't forget where we came from, and we don't forget that long narrative that's developed, and our particular insight into both the blessings but also the more painful parts of our past."

What if the president might be seen and *felt* discussing, clearly, in one place, at one time, the painful parts of our past: that slavery was wrong, that Jim Crow was wrong, that white terror was wrong, and that those Americans who held, recalcitrantly, to the murderous pledge of allegiance to a kingdom of hate under a banner of bigotry were just wrong? And what if he managed to accentuate the blessings that occasionally, through tragedy, come to clarify the nation's moral trust with its darker kin even as their suffering redeems the nation's political ideals? That might be a grace too amazing to imagine.

8

AMAZING GRACE

Obama's African American Theology

THE LAST WEEK OF JUNE 2015 HAD BEEN THE GREATEST WEEK OF Obama's presidency, and one of the greatest weeks any president had ever had.[1] The Supreme Court had delivered surprising victories for the president by upholding Obamacare's nationwide insurance subsidies on Thursday, and legalizing same-sex marriage across America the next day. Although it wasn't as loudly touted, the nation's highest court on Thursday also upheld a central legal argument put forth by the Obama administration: that claims of racial discrimination in housing shouldn't be limited by questions of intent but should be determined by discriminatory effect. Obama had begun the week with a big victory on his Pacific Rim trade deal by receiving a green light from the Senate to negotiate trade with eleven countries.

And then on Friday afternoon, Barack Obama delivered the eulogy for fallen South Carolina state senator and Emanuel AME pastor Reverend Clementa Pinckney — arguably the most crucial speech of his presidency. The nation had been shocked by Dylann Roof's massacre of nine souls at prayer at Emanuel. The president stepped into the pulpit to celebrate a martyr for black freedom, which has always meant, but never more clearly than now, the freedom of the nation to be its best.

Obama knew the minister, but not well, a fact that had moral utility: Pinckney was a proxy for all those who had lost their lives in the recent siege of racial terror that was sweeping the nation. Roof claimed in his sick manifesto that black people were taking over, a delusion easily rebutted on the same Internet that fed the gunman's twisted logic. No single person better embodied black progress, and therefore scared white terrorists more, than Barack Obama. Could it be that unarmed blacks who were dying across the nation were urban proxies for the black presidency and the change it had brought? Those who can't aim a gun at Obama take whatever black lives they can. Roof is not, therefore, a lone wolf. A better way of saying this is that calling him a lone wolf hardly denies the hatred of the philosophical pack from which he separated; the evil he reflects is deeply entrenched in our culture. The banner he killed under did not go away when the Confederate flag—which should have come down long ago—was removed.[2]

When Obama stood in the pulpit as president, he bore a guilt for which he was in no way individually responsible; it was instead a symbolic guilt as the shining emblem of black mobility that the killer found so offensive. It is a guilt that very few human beings can ever fully know or understand: Martin Luther King Jr. was wracked with guilt for being the centerpiece of a freedom movement that so many others participated in and died for. It was not simply a matter of knowing that many others deserved credit. It was understanding that his actions provoked, deliberately at times, the established order to deadly response: fire hoses washing away black bodies, police dogs tearing into black flesh, billy clubs lowered on black heads, bullets unloaded in black backs. Of course King was not responsible for a single death as leader of the movement; that responsibility lay in the belly of white terror that he agitated, and which spit up those black bodies in retching compensation for the pressure they put on the anatomy of white privilege. But King still felt guilt despite knowing that what he did was necessary if black people were to be free in a racist society that would deny their freedom as long as they were afraid to demand it or die for it. For King it wasn't merely a matter of racial guilt, either, but a theological one as well. Every Christian minister is a stand-in for God, who, as Christ, grappled with death and bore the weight of sin, of human guilt, on his shoulders. The act of preaching is the ritual reenactment of that sacrifice, and of symbolically taking on guilt. Obama bore both theological and political guilt that day when he climbed the pulpit as our nation's First Preacher.

Obama's body, again, was at a crossroads. He had often spoken of the benefits of his biracial biography; it was needed now more than ever, for the black and white elements that went into his making were at loggerheads, incessantly. Had his presidency put an end to racism or reignited its burning terror? Roof, of course, gave his answer, mad, literally, that black folk seem always to talk of race while whites hardly speak of it at all. Of course black folk spoke of the pain in their gut, the burr under their saddle, the weight around their neck, the invisible cloud that poured rain on their slow and forever-delayed parade to parity. Even whites who are not racists wonder the same thing: Why are blacks obsessed with race; why can't they stop navel-gazing when the world lies before them for the taking; instead of moaning the blues, why can't they make an honest go of it like whites had to?

Obama had tapped some of this sentiment in his speeches, especially his famed race address in Philadelphia. But now was not the time to lay into blacks, but to back them; an unforeseen opening had arrived when Governor Haley and Senator Graham of South Carolina both said that the Confederate flag should no longer wave its despotic message over their State House grounds. Like Obama, Haley and Graham bore a kind of symbolic guilt, too, for having vigorously defended a symbol of hate that masqueraded instead as a sign of southern heritage. Haley and Graham thus expiated their guilt, and the guilt of their fellow South Carolinians, by vigorously embracing an interpretation of the flag they would have disputed the day before Roof's racist rampage.

If ever there was doubt that Obama is the celebrated compromise between black assertion and white resistance, moments like this confirmed his value on either side of the divide, which, inevitably, left both sides hungry for more. He could never be white enough, given his brown skin; and his blackness had always to be contained, circumspect, signified, radiated, implicit—always threatening to flare at the wrong moment. Calling a white cop stupid. Saying a black boy in Florida killed by a rogue neighborhood watchman could have been his son. Obama had always to field demands from some blacks to be blacker, and the wish of many whites to whitewash the story of American race and politics. He was shadowed between the ideals and the realities of race and identity in America. But he had made it a productive tension, one that put him in office and thus increased the pressure on him to reconcile the two—a challenge he had often shirked, one from which he begged relief, one he sometimes denied being aware of. Obama had often had to be both Christ and Peter: the evidence of the resurrection of racial progress to come, and the denial of the suffering blackness through which it might emerge.

If Haley and Graham had to confess their guilt about the flag, Obama had fences to mend, too, for having dispensed to black people too much tough love and not enough tender loving care. It was important for the nation to see him loving black people without hesitation or apology. There is political import to such displays, for they highlight elements of black culture that are usually ignored until crisis comes. Just as Obama took pains to insist that most Americans are decent people, it is equally true that most black folk are loving and forgiving people. It may be true that those qualities—like survivors forgiving a white racist his evil deed even before their loved ones found their final resting place—shine brightest in catastrophe.

But they beam, too, in everyday acts of valor, particularly in what blacks refuse to do: hate all police because of instances of egregious brutality, or make violent attempts to undo a criminal justice system that unfairly punishes black people. It is a wonder that black goodness, and sanity, survive the hidden injuries of race: the daily denials of opportunities; the withholding of resources, goods, and services; and the relentless assaults, both subtle and vocal, on the humanity of black folk. Obama knew better and should have said so more often, more publicly, more loudly. Now was his chance to broadcast to the world the beauty of black humanity.

When Obama stepped into the pulpit, he was greeted with organ chords and drumrolls, and a sea of royal purple behind him as the majesty of the AME bishopric engulfed him. The Baptist and Methodist churches, with their loose organizational structures and their less formal educational standards, had historically been more open to black slaves and their style of worship.[3] When they left to form their own colored congregations, black folk sought to expand their biblical literacy and to deepen their mastery of rhetoric. Among those on the stage behind Obama was Vashti McKenzie, the first female bishop of the AME Church and one of the most gifted preachers in the nation. In the vast congregation of six thousand stretched out before him at the College of Charleston were Jesse Jackson and Al Sharpton, past masters of black sacred speech. Obama was surrounded by black rhetorical genius and the greatness of black music, and had every reason to tap both veins in his performance.

"Giving all praise and honor to God," Obama began. Right off the bat he was signaling that he would not operate, at least not exclusively, as a politician, or even as president, but as preacher, with the phrase that accompanies many testimonies, speeches, and sermons in black churches — with a slight emendation: "Giving all praise and honor to God, who is the head of my life."[4] Obama need not go that far, to claim God as the head of his life, since that dependent clause might signal for some a personal claim that would make him vulnerable to rebuff — "Really? God is the head of your life as you conduct war and send drones?"—or unkind parody. The phrase as it stood drew applause and verbal affirmation from the audience.

"The Bible," Obama said next, leaving no doubt as to his rhetorical anchor, following the Spirit with the Word of God, "calls us to hope. To persevere, and have faith in things not seen." He didn't have to mention that it was the same kind of hope he often spoke of, an audacious hope, a hope for which he named his second book as a way to capture his political vision. Obama read his scripture, as preachers are wont to do, except, unlike most of them, he didn't tell us the source of his words, which in this case are from the eleventh chapter of the New Testament Book of Hebrews, a book that famously details the substance of faith and hope. We sensed there was something monumental about to happen, since only the epic can match the tragedy of what had occurred; it was being presaged in his announcing "scripture," but not a particular scripture. The universal was hovering low, about to claim the eulogy as a conduit for the sort of truth that no one scripture can contain, but which reflects the whole of the book itself—the very idea of God's word.

"We are here today to remember a man of God who lived by faith. A man who believed in things not seen. A man who believed there were better days ahead, off in the distance. A man of service who persevered, knowing full well he would not receive all those things he was promised, because he believed his efforts would deliver a better life for those who followed."

After acknowledging Pinckney's widow and daughters (how personal and painful that must have been, as Obama—in the presence of his own beautiful wife, Michelle, with their two precious daughters, Sasha and Malia, back home in Washington—saw before him the reflection of his own family and possible fate), Obama foreshadowed his theme of grace: "The first thing I noticed was his graciousness, his smile, his reassuring baritone, his deceptive sense of humor—all qualities that helped him wear so effortlessly a heavy burden of expectation." Could not these words be spoken of the eulogist himself? His voice may have been an octave higher, his skin less chocolate, but they shared the same smile and humor. Black men so bright and accomplished so rarely offer love and recognition to one another in public, except, perhaps, on a hardwood court or at a music or movie awards show, and it is a shame that it is too often in death that compliments are bestowed.

Obama saluted Pinckney's preacherly pedigree, telling us that he came from a long line of ministers who "spread God's word, a family of protesters who sowed change to expand voting rights and desegregate the South. Clem heard their instruction, and he did not forsake their teaching." Obama tied together Pinckney's political and religious vocations, calling him an "anointed" man. Obama, of course, had grown to appreciate great preaching under the tutelage of Jeremiah Wright; this was an ode, indirectly, likely unintentionally, but no less effectively, to the spiritual side of Wright, the man who shaped Obama's religious identity, a fact that gained little notice in their bitter contretemps. Obama spoke of Pinckney's political work in representing "a sprawling swath of the Lowcountry, a place that has long been one of the most neglected in America. A place still wracked by poverty and inadequate schools; a place where children can still go hungry and the sick can go without treatment. A place that needed somebody like Clem."

What he didn't say, perhaps felt no need to say in light of the massacre, is that the problems of blacks in South Carolina run far deeper than bad schools, hunger, poverty, and poor health care. Black folks had waged perennial war against white supremacy from the day they set foot on South Carolina soil.[5] Blacks had been rebelling against, and running away from, the brutal oppression of slavery and its punishing subtropical clime at least since September 9, 1739, the day of the infamous Stono Rebellion near Charleston, where forty-four black people, and twenty-one whites, were killed. It was the largest slave rebellion in the history of the British mainland colonies, though of course there were larger insurrections in the Caribbean and South America. The blacks in South Carolina were trying to get to Florida, which was then under Spanish rule; the British controlled

Carolina. The king of Spain had issued an edict announcing that any slave who made it across St. Mary's River in Florida would be free. Black folk made every effort on land and sea to get to St. Mary's and secure their freedom. Those who made it created the first free black town in North America, Fort Mose, in Florida.

It should be remembered that 48 percent of all of the black people who entered the colonies, and the United States after 1776, until 1808, when the slave trade was outlawed, came to America through Charleston. It was so identifiably black that even the white people called it "Negro Country."⁶ There is little wonder that Charleston is such a contested site for racist diehards and is ground zero in the history of tortured race relations in America. This history cannot be ignored in grappling with the Confederate flag and its relationship to the animus that flashed in deadly manner at Emanuel. Antiblack racism flows like a vicious current under the floorboards of American history; but under the floorboards of South Carolina, it rushes like a flood. Black people fought back in South Carolina; they risked their lives, and took the lives of their oppressors, to be free, and that terror — festering for nearly three hundred years — is just one of the motivations for the virulent racism that flowed through Dylann Roof into the lives of Charleston and indeed the nation.

The theme of grace played on Obama's lips, especially as he contrasted grace to fallen humanity; the killer, Obama declared, was blind to grace in all the ways it shone during and after his fatal act.

> Blinded by hatred, the alleged killer could not see the grace surrounding Reverend Pinckney and that Bible study group — the light of love that shone as they opened the church doors and invited a stranger to join in their prayer circle. The alleged killer could have never anticipated the way the families of the fallen would respond when they saw him in court — in the midst of unspeakable grief, with words of forgiveness. He couldn't imagine that.

The alleged killer could not imagine how the city of Charleston, under the good and wise leadership of Mayor Riley [,] . . . how the state of South Carolina, how the United States of America would respond—not merely with revulsion at his evil act, but with big-hearted generosity and, more importantly, with a thoughtful introspection and self-examination that we so rarely see in public life.

Obama explicitly embraced his theme as he acknowledged that for the "whole week, I've been reflecting on this idea of grace." Obama spoke of the grace of the families left behind by the massacre, and said that grace was a topic Pinckney had often preached about. Obama recited lines from one of his favorite hymns: "Amazing grace, how sweet the sound that saved a wretch like me. I once was lost, but now I'm found; was blind but now I see." Obama brilliantly played on the blindness that plagues the American soul—blindness to the pain unfurled in waving the Confederate flag; blindness to how the harm of the past, in slavery and Jim Crow, caught up to the present in mass incarceration, tension and mistrust between cops and blacks, and the attack on voting rights—and how, with incredible grace, with God's grace, black folk overcame.

Like any good preacher, Obama had a working theological definition of grace as he encouraged the congregation to forge ahead and find the meaning of God's grace in their hearts and in neighborhoods across the land:

According to the Christian tradition, grace is not earned. Grace is not merited. It's not something we deserve. Rather, grace is the free and benevolent favor of God as manifested in the salvation of sinners and the bestowal of blessings. Grace. As a nation, out of this terrible tragedy, God has visited grace upon us, for he has allowed us to see where we've been blind. He has given us the chance, where we've been lost, to find our best selves. We may not have earned it, this grace, with our rancor and complacency, and short-sightedness and fear of each other—but we got it all the same. He gave it to us anyway. He's once more given us grace. But it is up to us now to make the most of it, to receive it with gratitude, and to prove ourselves worthy of this gift.

It's a dangerous prospect for a professional politician to wade into theological complexities and to offer his view of faith, but Obama garnered confidence in the preaching moment. He shows, too, the political utility of grace. There is a subtle rebuke to political self-portraits of romantic, rugged individualism that refashioned a do-it-yourself pragmatism into a mythology of the self-made man. We are all indebted, the president reminds us, to a grace we didn't earn—and of course the nation has also benefited from so much black labor and black grace that it didn't earn. Obama had earlier, in the 2012 campaign, during a stump speech in Virginia, got in trouble with conservatives when he claimed that "if you've got a business"—in the line that got yanked out of context and repeated—"you didn't build that." But the broader context and true meaning were deliberately missed:

There are a lot of wealthy, successful Americans who agree with me—because they want to give something back . . . [I]f you've been successful, you didn't get there on your own . . . I'm always struck by people who think, well, it must be because I was just so smart. There are a lot of smart people out there. It must be because I worked harder than everybody else. Let me tell you something—there are a whole bunch of hardworking people out there.

If you were successful, somebody along the line gave you some help. There was a great teacher somewhere in your life. Somebody helped to create this unbelievable American system that we have that allowed you to thrive. Somebody invested in roads and bridges. If you've got a business—you didn't build that. Somebody else made that happen . . .

The point is, is that when we succeed, we succeed because of our individual initiative, but also because we do things together . . . We rise or fall together as one nation and as one people.[7]

In the context of the funeral, the reminder of grace had a political and racial resonance that was hard to miss.

If Obama had been, before, hesitant to embrace his blackness, and to call it by its name, allow it to name him, and to use it as a lens on America, the hesitation melted this day. In a remarkable passage Obama argues that America has been blind to how the past colors our present racial moment and therefore blind to how we have missed out on God's grace:

Perhaps we see that now. Perhaps this tragedy causes us to ask some tough questions about how we can permit so many of our children to languish in poverty, or attend dilapidated schools, or grow up without prospects for a job or for a career. Perhaps it causes us to examine what we're doing to cause some of our children to hate. Perhaps it softens hearts towards those lost young men, tens and tens of thousands caught up in the criminal justice system and leads us to make sure that that system is not infected with bias; that we embrace changes in how we train and equip our police so that the bonds of trust between law enforcement and the communities they serve make us all safer and more secure. Maybe we now realize the way racial bias can infect us even when we don't realize it, so that we're guarding against not just racial slurs, but we're also guarding against the subtle impulse to call Johnny back for a job interview but not Jamal. So that we search our hearts when we consider laws to make it harder for some of our fellow citizens to vote. By recognizing our common human-ity[,] by treating every child as important, regardless of the color of their skin or the station into which they were born, and to do what's necessary to make opportunity real for every Ameri-can—by doing that, we express God's grace.

By insisting that Americans not harm black people by doing these unjust things to them, and telling us that the will to name, and undo, harm is an expression of God's grace, Obama goes far beyond political arguments about resources and links the doing of justice to the moral order of the universe. If in the past Obama lagged far behind in insisting on the dignity of black identity—in acknowledging his own blackness and how it might have anything to do with how he thought or behaved as president—in his eulogy Obama leaped cosmic dimensions to compassionately embrace a broader, bigger, blacker notion of blackness than ever before. He was not merely preaching to the world, to the nation, to the congregation, or to the choir. He was also preaching to himself. As the Negro spiritual says, "Not my brother, nor my sister, but it's me, O Lord, standing in the need of prayer."

Obama had, before his endorsement of blackness, also punctured the self-image of the southern white who claimed that heritage, not hate, marched beneath the Confederate flag. Obama insisted that the flag represented a legacy of white supremacy that harmed black citizens,[8] and that the grace that we claimed in the religious world must extend to politics as well:

> For too long, we were blind to the pain that the Confederate flag stirred in too many of our citizens. It's true, a flag did not cause these murders. But as people from all walks of life, Republicans and Democrats, now acknowledge—including Governor Haley, whose recent eloquence on the subject is worthy of praise—as we all have to acknowledge, the flag has always represented more than just ancestral pride. For many, black and white, that flag was a reminder of systemic oppression and racial subjugation. We see that now.
>
> Removing the flag from this state's capitol would not be an act of political correctness; it would not be an insult to the valor of Confederate soldiers. It would simply be an acknowledgment that the cause for which they fought—the cause of slavery—was

wrong. The imposition of Jim Crow after the Civil War, the resistance to civil rights for all people was wrong. It would be one step in an honest accounting of America's history; a modest but meaningful balm for so many unhealed wounds. It would be an expression of the amazing changes that have transformed this state and this country for the better, because of the work of so many people of goodwill, people of all races striving to form a more perfect union. By taking down that flag, we express God's grace.

Of course, Obama didn't have the exegetical or expository room to tackle another prominent symbol that has also shielded a multitude of sins: the cross. In the same way that the Confederate flag has flown above bigotry, the cross has been raised, too, against black people in the name of white terror, and against women, and gay and lesbian folk, and against transgender people, too, in the name of sexual and moral purity. God's grace was manifest, too, in the mingling of two victories on the same day: the Supreme Court decision legalizing same-sex marriage, and the Confederate flag receiving its most potent symbolic lowering in Obama's speech. In the black religious setting where he metaphorically folded the Confederate flag, Obama let stand a contradiction he might have reasonably assailed on another occasion: how the black church often undercut its own glorious reach, its own universal sweep, by emulating the bigotry it despised, bigotry that had unjustly curtailed black life. Many black believers had recoiled when Obama came out in support of same-sex marriage, using scripture and tradition to hammer gays and lesbians the same way white bigots had used the Bible to strike against black humanity and civil rights. Many black believers even chafed at the suggestion of a parallel between race and sex, but of course, race hate and homophobia flow from the same river of repulsion. In truth, there were many confederacies of bigotry operating that day, not just the Confederacy; many flags were flying for blindness; many banners were hung and waved for disbelief in our essential togetherness and unity. Either all black lives mattered or none mattered at all.

Obama underscored the literally unbelievable humanity of black people who could, without the killer ever asking, forgive the man who murdered their kin in cold blood. Theirs was no offer of "cheap grace," as the theologian Dietrich Bonhoeffer phrased it, "the preaching of forgiveness without requiring repentance." Bonhoeffer counseled instead the embrace of costly grace, in which "the gospel . . . must be sought again and again."⁹ The forgiveness of the folk in Charleston was more than a singular moral act; it was, too, a gesture of theological preemption and political strategy—a refusal to define their lives by hate and a refusal to offer the killer the pleasure of the race war he desperately hoped to provoke. Obama wasn't simply an observer of the black theology at work in forgiveness; he was a participant as well when he argued that Roof didn't know that God was using him. That doesn't mean that God intended for Roof to do wrong; it means that God uses even Roof's bad actions for good purposes, purposes that were lost to Roof behind his veil of evil. Black believers never tire of quoting the scripture in Genesis 50:20 about the good outcome of Joseph being sold into slavery by his brothers, only to become minister of agriculture to save them and a nation from famine: "You intended to harm me, but God intended it for good to accomplish what is now being done, the saving of many lives." The death of nine valiant souls brought down a symbol of hate that has menaced millions of blacks for decades—and may open the way to a more vigorous challenge to the white supremacy for which that flag stands. If black Christians didn't believe this, they would lose their minds and their faith in the God who guided them to triumph over suffering.

Perhaps it was spontaneous reflection on the grace of God, and the grace of black folk to bring the Word to life in their willingness to forgive—and the grace of black worship, with its mighty songs and its eloquent words—that caused a hush in the president, a silence, a pause, twice, as he repeated near the end of his eulogy the words "Amazing grace." "If we can find that grace, anything is possible. If we can tap that grace, everything can change." And then several seconds of silence. "Amazing grace," the president said, and then again, shortly afterward, "Amazing grace." That was followed by a longer silence, thirteen seconds of it—the silence necessary for speech to sink in, for speech to rise in the first place, for speech to make meaning and make an impact, for words to reach their biggest targets, the hearts and minds and souls of those listening. Obama played the pauses as brilliantly as he had commanded the speech; his rhetoric had soared to heights, and now it sank to silence, but a pregnant, creative silence, a cessation of talking that focused the energy of the moment on what was not being said, on what was being thought silently in the minds of those listening, and, finally, what was being felt by those who savored the ecstasies of Obama's eulogy.

There was much to contemplate in the silence. How Obama had broken free of constraints and given voice to a blackness that now percolated through his rhetoric. How the president had become a preacher leading the nation to spiritual healing. How the words of a black man were being heard around the world because a white man had desperately wished to unleash tides of hate that would forever mute the black voice. How a black presidency had looked as if it were on its last legs when suddenly, fresh momentum turned the president from lame duck to rising phoenix. How all the promise of hope and change that his presidency had pointed to, but continually frustrated, now seemed possible again because he gained the courage to be at his blackest when he was at his best. How he was fully and proudly African American.

At that moment, in those pauses, speech seemed hardly enough. Isn't that why we have always relied on our greatest artists to search our truths and sing our lives in their words? Why we have embraced words that are sung so that we might draw nearer to a vision or feeling or wish or hope that could never be captured in prose? The rhythm of our creation is mirrored in the rhythms of the songs we create to praise our inspirations and, for many of us, the God who made us with nobody else in mind. Obama paused to drink in all of this; he paused the first time for focus, the second time for courage. He knew that as powerful and as eloquent as his words had been—preachers, even temporary ones, know when they strike rhetorical fire—a more lasting impression could be made if he gave of himself even more. If he slipped out from beneath his security blanket and wrapped his vulnerability around the church—and as president, he leads a congregation that spans the nation—then his parishioners, his constituents, his countrymen and women, might be reminded of the need to risk their identities, comforts, securities, and pride, and go out on a limb for somebody too.

One could glimpse his struggle to find the right key in those pauses, as he glanced down, seeming to ask himself if he dared follow where the Spirit led him.[10] He repeated the phrase "Amazing grace" twice, between pauses, to remind himself that it was God's grace that would help him enchant the nation. And then, without warning, without a musical safety net, on the high wire of live television, before an audience of millions around the world, he stepped out on the faith he had encouraged others to follow. The president crooned a post-linguistic celebration of the truths he had evoked in such a masterly way. He memorably condensed into song what he had said over the last forty minutes. Singing, after all, is never just about singing; it is also about melodies breaking forth on one's lips which rise from one's heart and soul. Singing in church ratifies with the gut what the head has decided is true.

A singing preacher is a reminder that the message of God is both said and sung. A singing president is even more profound: a man becoming spiritually transparent for the world to see and hear; a figure going where no executive order can rescue notes ill flung, where no pen can veto the legislation of verbal dissent. When Obama, after electrifying, productive silence, launched into the first words of "Amazing Grace," the bishops and ministers behind him leaped to their feet. He turned his head, slightly, to acknowledge their approval as they chimed in to help him finish the verse.

Obama stayed mostly on tune, though he fell flat, a flatness that was both the object and vehicle of the blues that black folk embraced.[11] As Obama finished the verse, he spoke again, for speaking after singing—especially if that singing already followed speech—has to be engaging, and the president didn't disappoint. He called out the names of those who died with Pinckney that fateful night. Obama ushered his song into speech, his words now humming with the slight tune and gentle vibrato of black sacred rhetoric—rhetoric that could at any moment erupt into music within the spoken words, an art that in the black church is called the chanted sermon, or, more colloquially, "the whoop."[12] As he called each name, the president reminded us that they all "found that grace," dramatizing how much more amazing grace was for having been found in the midst of terror and grief and heartbreak and death.

Obama ended his eulogy and reveled in the warm elation of the bishops, and in that moment, and in all that had preceded it for the last forty minutes, the promise of his black presidency beamed as brightly as it ever had.

★|||||★

ACKNOWLEDGMENTS

THIS BOOK BEGAN THE NIGHT BARACK OBAMA WON THE PRESI-dency on November 4, 2008, when I witnessed the nation's first black president embrace a version of American exceptionalism and echo the immortal Sam Cooke: "It's been a long time coming, but tonight, because of what we did on this day, in this election, at this defining moment, change has come to America."

I had first met Barack Obama when the two of us served on a black history panel in the suburbs of Chicago in the very early nine-ties. I made an impassioned plea for Obama's U.S. Senate candidacy in 2003 at a local Chicago black radio expo, and introduced him to the crowd to say a few words before my keynote address that after-noon. After his Senate victory, I'd see Obama here and there — at the annual Christmas party thrown by *Ebony* magazine owner Linda

Johnson Rice, on the Amtrak Acela train between D.C. and Philadelphia—and we'd visit and have cordial conversations.

After he announced his presidential bid, I served as an informal advisor and official surrogate for Obama, traveling to Iowa before the first caucus, to Florida a few days before the 2008 election, and many places in between, touting his virtues. I notified him directly when I made an exception to my rule as a surrogate never to criticize him publicly as I penned an essay for *Time* magazine that took issue with his attack on black fathers in his famous July 2008 Father's Day address at Chicago's Apostolic Church.

The day after Obama's pioneering win, I was slated to deliver the prestigious W.E.B. Du Bois lectures at Harvard University, where Henry Louis Gates Jr. had invited me to hold forth on a subject of my choosing. I settled on rap icon Jay Z and came up with an ambitious title to match his outsized talent—"From Homer to 'Hova: Hustling, Religion, and Guerilla Literacy in the Pavement Poetry of Jay Z." But Obama stole the show. I scrapped my lectures on Jay Z, and for the next three days I wrestled out loud, an exercise in improvisational cultural criticism, with Barack Obama's monumental victory and how it might shape our ideas of politics and color. I am grateful to TV One and to Professor Gates and the Harvard community for giving me this rewarding opportunity to work out my initial ideas about President Obama and race in America.

My Du Bois lectures became the basis for my "Barack Obama and Race" seminar at Georgetown University, which draws some of Georgetown's most driven students who yearn to explore the political and racial contexts of their nation's first black presidency.

My Obama seminar lies at the nexus of our nation's racial anxieties and confusions—and our possibilities too. My class is a challenging space. We wage war against our own fears and ignorance, chasing racial fantasies out of the rabbit holes of culture in which they have sought cowardly refuge. No meaning of race is safe from our scrutiny, no consequence of race is closed to our consideration, no cherished truth of tribe or tradition is immune to our rigorous doubt. We try as best we can to see and tell the truth about Obama: he deserves to go down as one of the most important and consequential presidents in our history, and yet, when it came to race, he often stumbled.

In November 2015, Obama declared, "I am very proud that my presidency can help to galvanize and mobilize America on behalf of issues of racial disparity and racial justice. But, I do so hoping that my successor, who's not African American, if he or she is not, that they'll be just as concerned as I am, because this is part of what it means to perfect our union." Obama made the statement as he sought to sign an executive order to combat job discrimination for ex-offenders. But neither his comments about race, nor his actions on ex-offenders (a disproportionate number of whom are black and brown) would have happened without significant social protest against racial injustice in the streets, and principled criticism of Obama from black quarters. My seminar has grappled with such untidy revelations, and I am grateful to my students for helping me to clarify my thinking about Obama and race over the last several years.

The Black Presidency is the elaboration of several years of teaching and reflects on the concerns and struggles of my students. I have also witnessed the great hunger beyond the ivory tower for the issues my students and I discuss every fall. As I discovered firsthand in esteemed lecture halls, prisons, public libraries, sanctuaries, and corporate boardrooms, my classroom is a vibrant microcosm of the concerns and struggles of the nation at large.

I want to thank the following people for supporting my efforts to understand and explain Obama over the years. Tanya McKinnon, my literary agent, is a towering intellectual presence whose relentless push for clarity and conceptual rigor made me come to terms with my thoughts in ways I wouldn't have otherwise. Thank you for being *sui generis,* Tanya. Deanne Urmy is a prodigiously gifted editor whose broad vision and brilliant imagining with me of this book have made it far better than it would have been without her help. Thank you so much, Deanne, for working your magic and making me part of the Houghton Mifflin Harcourt family.

I also want to thank Amanda Heller, who offered conscientious attention to the manuscript and made it much stronger. Thanks as well to Lori Glazer, director of publicity, Brian Moore (great jacket photo, sir!), Lisa Diercks (great layout, ma'am), Kelly Dubeau Smydra, Jill Lazer, Ayesha Mirza, Stephanie Kim, Barbara Jatkola, Donna Riggs, Giulana Fritz, and Jenny Xu. Nina Subin, thanks for a very nice author photograph on the dust jacket. And great thanks as well to Beth Burleigh Fuller, a wonderful production editor who helped me clean things up at the end!

Paul Farber, my former student and now a superb scholar and colleague, helped in the early days with research, as did the very talented Mishana Garschi in the later period of my work. I thank brilliant thinkers James Braxton Peterson, Salamishah Tillet, and Marc Lamont Hill for listening to my thoughts and offering me their own, and for their love and support as well.

I'm grateful to Phil Griffin, quick-witted MSNBC impresario and my brother from the Midwest, as well as to the sublime Yvette Miley for making me a viable member of the MSNBC family, where I got plenty of time to talk about politics and race on air. I greatly appreciate the generosity of my man Ed Schultz, and that of his former producer James Holm, and for wonderful shows hosted by media maven and brilliant force of nature Tamron Hall, sculptor of words Alex Wagner, journalist and edifying soldier of the Cross Martin Bashir, gifted scholar Melissa Harris-Perry, and Joy Reid, a formidable thinker and graceful writer whose countless conversations about Obama and race were very helpful. Big thanks to the Reverend Al Sharpton, on whose MSNBC show I vetted some of these ideas, and in whose debt I remain for his generous interview for this book—and for his brilliant and courageous leadership over the years.

Thanks as well to Eric Holder (and his equally gifted wife, Sharon Malone) for his great interview, and for his revival of the office of the Attorney General as a forum for racial justice and equality. And while she and her boss will surely disagree with some of what I say here, I'm grateful to Valerie Jarrett for arranging my interview with President Barack Obama—to whom I'm greatly indebted for an honest and insightful conversation for this book.

I am also grateful to my *New York Times* family, especially the very talented Sewell Chan; Jessica Lustig, my former student and now my uncommonly wise teacher in many ways; and Rachel Dry, my enormously gifted editor, who helped me understand what I thought about race and Obama. A big shout of gratitude as well to my *New Republic* family, including the illustrious trio of editor-in-chief Gabriel Snyder, features director Ted Ross, and senior editor Jamil Smith.

I'm grateful to be part of the Georgetown faculty and for the collegiality of my fellow faculty in Sociology and African American Studies. I'm thankful for John DeGioia—one of the great university presidents in America, an unassuming, genuinely humble educational leader, whose spirit and heart are in the right place—and for his extraordinarily generous support of my work.

I'm also grateful to Susan Taylor and Khephra Burns—aka "the Queen of Black America" and "Smooth," two noble embodiments of incredible black humanity and huge intelligence—for their unending well of friendship, love, and support. The same is true for my daughters of the heart Janaye and Janique Ingram, and Angela Rye—inspiring young leaders, thinkers, and activists. And huge thanks to Farah Jasmine Griffin and Obery Hendricks, a dynamic scholarly and intellectual duo, for their big brains and eloquent pens, for their love and support, and their timely discussions of politics and race.

Profound gratitude goes to my dear friend and brother Reverend Dr. Frederick Douglass Haynes III, one of the most gifted sacred rhetoricians on the globe, who discussed many of these ideas with me over the years, and who permitted me to share them with his congregation, the Friendship West Baptist Church in Dallas. And big thanks to my beloved pastor, Reverend Dr. Howard-John Wesley, of the Alfred Street Baptist Church in Alexandria, Virginia, a truly phenomenal preacher and exquisite wordsmith, who permitted me to try out these ideas on our congregation in a Sunday school setting. And deep appreciation to the remarkable Dean of Howard University's Rankin Chapel, Dr. Bernard Richardson, a world-class spiritual leader in the mold of Howard Thurman, who has permitted me to preach these ideas to the good folk of Howard over the last decade.

Finally, I am grateful to my family: my wonderful mother, Addie Mae Dyson, still holding strong at seventy-eight as our matriarch; my splendid brothers, Anthony, Brian, Gregory, and Everett (prisoner #212687—Godspeed and hurry home!); my supremely talented children, Michael II, Maisha, and Mwata, and my lovely daughter-in-law, Wanda; and my beautiful grandchildren, Layla, Mosi, and Maxem. And to my loving wife, Marcia, to whom this book is dedicated: I am grateful for your undying love, your profound commitment to our family, and for your genius as a thinker of deep thoughts and profound ideas. You are one of the most remarkable women on earth. Thank you for being our family's grace and glue.

★ ||||| ★

PRESIDENT OBAMA'S SPEECHES
AND STATEMENTS ON RACE

1. DNC keynote address ("There is not a white America . . ."), Boston, July 27, 2004

2. First Selma speech, Brown Chapel AME Church, Selma, Alabama, March 4, 2007

3. Speech at Essence Festival, July 5, 2007

4. "Yes We Can" campaign speech, New Hampshire primary, January 8, 2008

5. Campaign speech, Sumter, South Carolina, January 24, 2008

6. "A More Perfect Union," Philadelphia, March 18, 2008

7. Father's Day speech, Apostolic Church of God, Chicago, June 15, 2008

8. Nomination acceptance speech, DNC, Denver, August 28, 2008

9. First Inaugural Address, Washington, D.C., January 20, 2009

29. Speech before U.S. Conference of Mayors (mention of Charleston), San Francisco, June 19, 2015

30. Speech honoring LGBT Pride Month (heckler speech), June 24, 2015

31. Eulogy for the Honorable Reverend Clementa Pinckney, Charleston, South Carolina, June 26, 2015

32. Speech before the NAACP 2015 Convention on criminal justice reform, Philadelphia, July 15, 2015

33. Speech at Congressional Black Caucus Dinner, September 19, 2015.

34. Remarks at White House Criminal Justice Reform Panel (where Obama defends "Black Lives Matter" movement), October 22, 2015.

35. Speech to the International Association of Chiefs of Police, Chicago, Illinois, October 27, 2015.

36. Victory speech on becoming the 44th president of the United States, Grant Park, Chicago, Illinois, November 4, 2008.

37. "Ban the Box" speech on the Prisoner Re-Entry Initiative, Newark, New Jersey, November 2, 2015.

★ ||||||| ★

NOTES

Introduction: The Burden of Representation

1. Barack Obama, "A More Perfect Union," in *The Speech: Race and Barack Obama's "A More Perfect Union,"* ed. T. Denean Sharpley-Whiting (New York: Bloomsbury USA, 2009), p. 237; Condoleezza Rice, interview on *Face the Nation*, CBS, November 27, 2011, http://www.cbsnews.com/news/condi-rice-us-will-never-be-race-blind/.

2. Quoted in Henry Louis Gates Jr., *Thirteen Ways of Looking at a Black Man* (New York: Random House, 1997), p. 18.

3. Thomas Jefferson, *Notes on the State of Virginia*, annotated ed. (New York: Penguin Classics, 1998); James Baldwin, *The Fire Next Time* (1963; repr., New York: Vintage, 1993).

4. John Campbell, *The Iron Lady: Margaret Thatcher, from Grocer's Daughter to Prime Minister*, abridged ed. (New York: Penguin Books, 2011); Charles

Moore, *Margaret Thatcher: From Grantham to the Falklands* (New York: Alfred A. Knopf, 2013).

5. Moore, *Margaret Thatcher,* pp. 298–333; Stuart Hall, *The Hard Road to Renewal: Thatcherism and the Crisis of the Left* (London: Verso Press, 1988).

6. When Baroness Thatcher died in 2013, President Obama issued an official statement and noted her gender as a defining element of her legacy: "As a grocer's daughter who rose to become Britain's first female prime minister, she stands as an example to our daughters that there is no glass ceiling that can't be shattered." "Statement from the President on the Passing of Baroness Margaret Thatcher," April 8, 2013, https://www .whitehouse.gov/the-press-office/2013/04/08/statement-president -passing-baroness-margaret-thatcher.

7. Although Disraeli was baptized into the Church of England at the age of twelve, his Jewish heritage remained a central feature of his existence and identity. See Adam Kirsch, *Benjamin Disraeli* (New York: Schocken Books, 2008). Thanks to historian Gerald Horne for suggesting the parallel between Disraeli and Obama in a brief, serendipitous conversation in an airport.

8. Thomas J. Carty, *A Catholic in the White House? Religion, Politics, and John F. Kennedy's Presidential Campaign* (New York: Palgrave Macmillan, 2004). For a fascinating comparison of Obama and Kennedy, see Robert C. Smith, *John F. Kennedy, Barack Obama, and the Politics of Ethnic Incorporation and Avoidance* (Albany: State University of New York Press, 2013).

9. Geoffrey C. Ward, *Unforgivable Blackness: The Rise and Fall of Jack Johnson* (New York: Alfred A. Knopf, 2004).

10. Dewayne Wickham, *Bill Clinton and Black America* (New York: One World/Ballantine, 2002).

11. Toni Morrison, "The Talk of the Town: Comment," *The New Yorker,* October 5, 1998, pp. 31–32. Chris Rock said in an interview in the August 1998 issue of *Vanity Fair* that Clinton was "the first black president." He also said that Clinton was "the most scrutinized man in history, just as a black person would be. He spends a hundred dollar bill, they hold it up to the light." See Jonathan Tilove, "Before Bill Clinton Was the 'First Black President,'" Newhouse News Service, March 6, 2007, http:// jonathantilove.com/before-bill-clinton-was-the-first-black-president/. In 2008, in *Time* magazine, when asked if she regretted referring to Clinton as the first black president, Morrison said that people "misunder-

stood that phrase. I was deploring the way in which President Clinton was being treated, vis-à-vis the sex scandal that was surrounding him. I said he was being treated like a black on the street, already guilty, already a perp. I have no idea what his real instincts are, in terms of race." See Toni Morrison, "10 Questions for Toni Morrison, *Time,* May 7, 2008, http://content.time.com/time/magazine/article/0,9171,1738507,00 .html. Indeed, in *The New Yorker,* Morrison wrote: "Years ago . . . one heard the first murmurs: white skin notwithstanding, this is our first black President. Blacker than any actual black person who could ever be elected in our children's lifetime. After all, Clinton displays almost every trope of blackness: single-parent household, born poor, working-class, saxophone-playing, McDonald's-and-junk-food-loving boy from Arkansas." According to Morrison, Clinton's blackness became even clearer when "the President's body, his privacy, his unpoliced sexuality became the focus of the [impeachment] persecution." Morrison, "Talk of the Town," p. 32. During a 2008 Democratic presidential debate in South Carolina televised live on CNN, journalist Joe Johns asked Obama if Clinton was the first black president.

"Well, I think Bill Clinton did have an enormous affinity with the African-American community, and still does," Obama said. "And I think that's well earned . . . [O]ne of the things that I'm always inspired by — no, I'm — this I'm serious about. I'm always inspired by young men and women who grew up in the South when segregation was still taking place, when, you know, the transformations that are still incomplete but at least had begun had not yet begun. And to see [those] transformations in their own lives[,] I think that is powerful, and it is hopeful, because what it indicates is that people can change.

"And each successive generation can, you know, create a different vision of how, you know, we have to treat each other. And I think Bill Clinton embodies that. I think he deserves credit for that. Now, I haven't . . . I have to say that, you know, I would have to, you know, investigate more of Bill's dancing abilities. You know, and some of this other stuff before I accurately judge whether he was in fact a brother." Wolf Blitzer said, "Let's let Senator Clinton weigh in on that." Hillary Clinton then humorously retorted, "Well, I'm sure that can be arranged." "Part 3 of CNN Democratic Presidential Debate," January 21, 2008, http://www .cnn.com/2008/POLITICS/01/21/debate.transcript3/.

12. Kenneth O'Reilly, *Nixon's Piano: Presidents and Racial Politics from Washington to Clinton* (New York: Free Press, 1995); Manning Marable, *The Great Wells of Democracy: Reconstructing Race and Politics in the 21st Century* (New York: Basic Civitas Books, 2002), pp. 77–84.

13. President Clinton admitted, both in a foreword to a book on criminal justice and in a speech before the 2015 NAACP convention—the day after President Obama at the same convention offered his landmark speech denouncing mass incarceration—that his policies had been wrong and harmful. "Plainly, our nation has too many people in prison and for too long—we have overshot the mark. With just 5 percent of the world's population, we now have 25 percent of its prison population, and an emerging bipartisan consensus now understands the need to do better." Clinton also argued that it is "time to take a clear-eyed look at what worked, what didn't, and what produced unintended, long-lasting consequences." He said that "some are in prison who shouldn't be, others are in for too long, and without a plan to educate, train, and reintegrate them into our communities, we all suffer." See "William J. Clinton: Foreword," April 27, 2015, https://www.brennancenter.org/analysis/foreword (from the book *Solutions: American Leaders Speak Out on Criminal Justice*, ed. Inimai Chettiar and Michael Waldman [New York: Brennan Center for Justice, 2015]). In his 2015 NAACP speech, Clinton conceded his error as president: "Yesterday, the president spoke a long time and very well on criminal justice reform. But I want to say a few words about it. Because I signed a bill that made the problem worse and I want to admit it." See Eric Levitz, "Bill Clinton Admits His Crime Law Made Mass Incarceration 'Worse,'" MSNBC.com, July 15, 2015, http://www.msnbc.com/msnbc/clinton-admits-his-crime-bill-made-mass-incarceration-worse. For the deleterious (racial) consequences of welfare reform, see, by Peter Edelman (who resigned as the assistant secretary for planning and evaluation at the Department of Health and Human Services in September 1996 in protest of Clinton's signing the welfare reform bill), "The Worst Thing Bill Clinton Has Done," *The Atlantic*, March 1997, http://www.theatlantic.com/magazine/archive/1997/03/the-worst-thing-bill-clinton-has-done/376797/. Also see Dylan Matthews, "Welfare Reform Took People Off the Rolls. It Might Have Also Shortened Their Lives," *Washington Post*, June 18, 2013, http://www.washingtonpost.com/blogs/wonkblog/wp/2013/06/18

/welfare-reform-took-people-off-the-rolls-it-might-have-also-short
ened-their-; Zenthia Prince, "Welfare Reform Garnered for Black
Women a Hard Time and a Bad Name," *Afro,* March 18, 2015, http://
www.afro.com/welfare-reform-garnered-for-black-women-a-hard
-time-and-a-bad-name/; and Bryce Covert, "Clinton Touts Welfare Reform. Here's How It Failed," *The Nation,* September 6, 2012, http://
www.thenation.com/article/clinton-touts-welfare-reform-heres-how
-it-failed/.

1. How to Be a Black President: "I Can't Sound Like Martin"

1. Associated Press, "Jackson Calls for War on Poverty; Hart Raps Mon-
    dale's Money Sources," *Sarasota Herald-Tribune,* April 23, 1984.

2. Marshall Frady, *Jesse: The Life and Pilgrimage of Jesse Jackson* (1996; repr.,
    New York: Simon & Schuster, 2006).

3. In his autobiography *An Easy Burden: The Civil Rights Movement and the
    Transformation of America,* foreword by Quincy Jones (1996; repr., Waco:
    Baylor University Press, 2008), p. 285, Andrew Young recounts this col-
    orful encounter. Referring to the other movement leaders, Young told
    King, "'Listen, Martin, I'm sick of being the bad guy; if they're such
    "geniuses" I'm tired of arguing with them all the time.' This really made
    Martin angry. 'I *depend* on you to bring a certain kind of common sense
    to staff meetings, and you know it,' he said. 'Now, if you decide you
    are going to start playing games, I don't see why I need you. I need you
    to take as conservative a position as possible, then I can have plenty of
    room to come down in the middle where I want to.'"

4. August Meier, "The Conservative Militant," in *Martin Luther King, Jr.: A
    Profile,* ed. C. Eric Lincoln, rev. ed. (New York: Hill & Wang, 1985), pp.
    144–156.

5. Republican New York congressman Pete King made this claim to the
    Today show's Matt Lauer in 2009. See David Edwards, "GOP Lawmaker:
    Obama Most Threatened President Ever," Alternet, 2009, http://
    www.alternet.org/rss/breaking_news/98972/gop_lawmaker%3A
    _obama_most_threatened_president_ever. Also see therehastobe
    away, "President Barack Obama Is the Most Threatened President in
    History," *Daily Kos,* November 25, 2012, http://www.dailykos.com
    /story/2012/11/26/1164628/--President-Barack-Obama-Is-the-Most

-Threatened-President-In-History; and Nathaniel Patterson, "The Most Threatened President in History," *Reader Supported News*, November 27, 2012, http://readersupportednews.org/news-section2/318-66/14744 -focus-the-most-threatened-president-in-history.

6. Geoffrey R. Stone, "Obama Africanus the First," *Huffington Post*, December 6, 2014, http://www.huffingtonpost.com/geoffrey-r-stone/obama -africanus-the-first_b_6282036.html.

7. Peter Wallsten, "Obama Struggles to Balance African Americans' Hopes with Country's as a Whole," *Washington Post*, October 28, 2012, http://www.washingtonpost.com/politics/decision2012/obama-after -making-history-has-faced-a-high-wire-on-racial-issues/2012/10/28/ d8e25ff4-1939-11e2-bd10-5ff056538b7c_story.html.

8. Barack Obama, "Remarks by the President at the 50th Anniversary of the Selma to Montgomery Marches," March 7, 2015, https://www .whitehouse.gov/the-press-office/2015/03/07/remarks-president-50th -anniversary-selma-montgomery-marches.

9. See Alvin Benn, "Lafayette on Stopping Disruption: 'Let President Speak," *Montgomery Advertiser*, March 8, 2015, http://www.montgomery advertiser.com/story/news/local/selma50/2015/03/08/lafayette-stop ping-disruption-let-president-speak/24600725/.

10. Barack Obama, "Selma Voting Rights March Commemoration Speech," Brown Chapel AME Church, March 4, 2007, http://www.american rhetoric.com/speeches/barackobama/barackobamabrownchapel .htm.

11. Alex Halperin, "Nastiest Conservative Responses to Obama's Trayvon Speech," *Salon*, July 19, 2013, http://www.salon.com/2013/07/19/best _of_the_worst_obamas_trayvon_speech/.

12. For a different take on the notion of implicit racial agreements between Obama and whites, see black conservative author Shelby Steele's argument about racial "bargainers" like Obama, who strike an agreement not to speak of race if whites agree not to remind them of their blackness, and racial "challengers"—blacks who accuse whites of being racist and then require them to absolve themselves of the charge by supporting affirmative action and cultural diversity—in *A Bound Man: Why We Are Excited About Obama and Why He Can't Win* (2008; repr., New York: Free Press, 2014). The fact that Obama won, and won again, suggests that Steele's core argument, first published as an essay in 2007 and premised

on Obama's inevitable loss at the polls, certainly possessed some insight but was fundamentally wrong then, and is more wrong now.

13. Barack Obama, "Statement by the President," July 14, 2013, https://www .whitehouse.gov/the-press-office/2013/07/14/statement-president. Obama's brief and dispassionate statement included the requisite nod to law and order: "I know this case has elicited strong passions. And in the wake of the verdict, I know those passions may be running even higher. But we are a nation of laws, and a jury has spoken. I now ask every American to respect the call for calm reflection from two parents who lost their young son."

14. Barack Obama, "Remarks by the President on Trayvon Martin," James S. Brady Briefing Room, the White House, July 19, 2013, https://www .whitehouse.gov/the-press-office/2013/07/19/remarks-president -trayvon-martin.

15. Barack Obama, "Remarks by the President on 'My Brother's Keeper' Initiative," February 27, 2104, https://www.whitehouse.gov/the-press -office/2014/02/27/remarks-president-my-brothers-keeper-initiative.

16. Clarence Page, "Millennials Are Just as Prejudiced as Their Parents," *Chicago Tribune,* March 17, 2015. Despite a 2010 Pew Research Report which maintains that more than two decades of research confirms that millennials are more tolerant than earlier generations, analysts like Spencer Piston, an assistant professor of political science at Syracuse University, argues that a closer examination of the data reveals persistent bias. Piston "examined the 2012 American National Election Studies racial stereotype battery, in which survey respondents are asked to rate whites, African-Americans, Hispanics, and Asians according to how hard-working or intelligent they are, and found something startling: Younger (under-30) whites are just as likely as older ones to view whites as more intelligent and harder-working than African-Americans (among the older cohort, 64 percent felt this way, and among the younger cohort the number was 61 percent — not a statistically significant difference). 'White millennials appear to be no less prejudiced than the rest of the white population,' Piston told Science of Us in an email, 'at least using this dataset and this measure of prejudice.'" See Sean McElwee, "Milennials Are Less Racially Tolerant Than You Think," *Science of Us,* January 8, 2015, http://nymag.com/scienceofus/2015/01/millennials -are-less-tolerant-than-you-think.html.

17. "Full Transcript: Obama's Remarks on Ferguson, Mo. and Iraq," *Washington Post*, August 18, 2104, http://www.washingtonpost.com /politics/running-transcript-obamas-remarks-on-ferguson-mo-and -iraq/2014/08/18/ed29d07a-2713-11e4-86ca-6f03cbd15c1a_story.html.

18. On January 20, 2009, the night of Obama's first inauguration, several Republican leaders in Congress — along with former House speaker Newt Gingrich, conservative journalist Fred Barnes, and conservative communications specialist Frank Luntz — gathered in the Caucus Room steakhouse in Washington, D.C., to plot, among the fifteen white men assembled, to undermine and disrupt government under an Obama administration and make him a one-term president. "You will remember this day," Speaker Gingrich said. "You'll remember this as the day the seeds of 2012 were sown." For a discussion of their meeting and its aims, see Robert Draper, *Do Not Ask What Good We Do: Inside the U.S. House of Representatives* (New York: Free Press, 2012), esp. pp. xv–xxii.

19. Cited in Michael Eric Dyson, *April 4, 1968: Martin Luther King, Jr.'s Death and How It Changed America* (New York: Basic Civitas Books, 2008), pp. 224–225.

20. Keli Goff, "Could Gay Marriage Spur Black Voter Drop?," *The Root*, September 17, 2012, http://www.theroot.com/blogs/blogging_the _beltway/2012/09/emanuel_cleaver_on_why_gay_marriage_could _cost_obama_black_votes.html.

21. I am not arguing that these are the only kinds, or groups, of black people to criticize Obama in some measure. For instance, Harvard professor Randall Kennedy offers in his excellent book *The Persistence of the Color Line: Racial Politics and the Obama Presidency* (New York: Pantheon, 2011) insightful criticism of Obama — his views on same-sex marriage, before he changed course; his "excessive cautiousness" on a range of issues; and his failure to stand up for the virtues of liberalism on the Supreme Court in the same way George Bush did for conservative justices during his tenure. But he is not visibly or vocally associated with a camp that was especially critical of Obama. Neither is Joy Reid, national correspondent for MSNBC and author of the very fine *Fracture: Barack Obama, the Clintons, and the Racial Divide* (New York: HarperCollins, 2015), an instant classic of political journalism that tackles the use of race against, and by, candidate Obama, and also features, besides a superb chronicle of the noxious racial forces Obama and his administra-

tion have confronted, unflinching engagement with Obama's failure, for instance, to speak honestly about police brutality after Ferguson, and his relentless chiding of black America. I am simply arguing that these individuals and groups named here are among the most visible and vocal critics of Obama and can be easily identified as such.

22. "President Obama, Congressional Black Caucus: No Meeting in 675 Days," Politics365.com, http://politic365.com/2013/03/18/president -obama-congressional-black-caucus-no-meeting-in-675-days/.

23. April D. Ryan, "CBC Chair Marcia L. Fudge Sends Letter to President Obama over Lack of African American Cabinet Appointments," March 11, 2013, http://aprildryan.com/2013/03/11/cbc-chair-marcia-l-fudge-sends -letter-to-president-obama-over-lack-of-african-american-cabinet -appointments/.

24. Evan McMorris-Santoro, "After Complaints About Diversity in 2nd Term Appointments, Congressional Black Caucus Thanks Obama," *BuzzFeed*, July 9, 2013, http://www.buzzfeed.com/evanmcsan/after-complaints -about-diversity-in-2nd-term-appointments-co#.reB9WOoO.

25. Jeff Johnson, "Rep. Waters to Black Voters: 'Unleash Us' on Obama," *The Grio*, August 17, 2011, http://thegrio.com/2011/08/17/frustration -boils-over-at-black-caucus-detroit-town-hall/. For a spirited defense of Waters, see Marcia Dyson, "Take Me to the Waters," *Huffington Post*, August 19, 2011, http://www.huffingtonpost.com/marcia-dyson/maxine -waters-obama_b_931276.html. Dyson argued that it "is high time for black folk to stop beating down on those of our race who dare lift their voices to offer constructive challenges to the White House. I don't mean personal or mean-spirited attacks; there's no place for that in our public discourse. I'm talking about well-reasoned and principled objections to this policy or that one, or the failure to lead in a political direction that benefits our communities. The stakes are high and the situation is critical in black neighborhoods and households across the land. We don't have time for bowing down at the throne of unbroken racial solidarity when our children are suffering, our elders are vulnerable, and our poor are teetering on the brink of economic and social disaster."

26. David Goldstein, "Black Caucus Treads Line Between Criticizing, Supporting Obama," *McClatchyDC*, http://www.mcclatchydc.com/news /politics-government/article24699058.html.

27. Kevin Johnson, "A President for Everyone, Except Black People,"

Philadelphia Tribune, April 14, 2013, http://www.phillytrib.com/news/a-president-for-everyone-except-black-people/article_164f06d9-abf2-5f29-a531-10ff57edf5f2.html.

28. Fredrick Harris, *The Price of the Ticket: Barack Obama and the Rise and Decline of Black Politics* (New York: Oxford University Press, 2012).

29. Fredrick Harris, "Still Waiting for Our First Black President," *Washington Post,* June 1, 2012, http://www.washingtonpost.com/opinions/still-waiting-for-our-first-black-president/2012/06/01/gJQARsT16U_story.html. For the "first gay president" claim, see Andrew Sullivan, *Newsweek,* May 21, 2012, http://www.newsweek.com/andrew-sullivan-barack-obamas-gay-marriage-evolution-65067.

30. Brittney Cooper, "Stop Poisoning the Race Debate: How 'Respectability Politics' Rears Its Ugly Head — Again," *Salon,* March 18, 2015, http://www.salon.com/2015/03/18/stop_poisoning_the_race_debate_how_respectability_politics_rears_its_ugly_head_again/, and "America's 'Black Body' Reality: How Selma, 'Scandal' & Ferguson Reveal an Ugly Truth," *Salon,* March 11, 2015, http://www.salon.com/2015/03/11/black_bodies_are_still_white_property_what_selma_scandal_ferguson_reveal_about_america/.

31. Brittney Cooper, "Black Girls' Zero-Sum Struggle: Why We Lose When Black Boys Dominate the Discourse," *Salon,* March 6, 2014, http://www.salon.com/2014/03/06/black_girls_zero_sum_struggle_why_we_lose_when_black_men_dominate_the_discourse/.

32. Ibid.

33. Ibid.

34. Brittney Cooper, "'Not Going to Lie Down and Take It': Black Women Are Being Overlooked by This President," *Salon,* June 17, 2014, http://www.salon.com/2014/06/17/not_going_to_lie_down_and_take_it_black_women_are_being_overlooked_by_this_president/.

35. Ibid.

36. Kimberlé Williams Crenshaw, "The Girls Obama Forgot: My Brother's Keeper Ignores Young Black Women," *New York Times,* July 29, 2014, http://www.nytimes.com/2014/07/30/opinion/Kimberl-Williams-Crenshaw-My-Brothers-Keeper-Ignores-Young-Black-Women.html?_r=0.

37. Ibid.

38. Glen Ford, "2007: The Year of Black 'Media Leaders' — Especially

Obama," *Black Agenda Report*, January 2, 2008, http://www.blackagenda
report.com/content/2007-year-black-'media-leaders'—especially
-obama.

39. Bruce Dixon, "Tired Old So-Called Leftists Give Same Old Excuses
for Supporting Obama in 2012," *Black Agenda Report*, August 15, 2012,
http://blackagendareport.com/content/tired-old-so-called-leftists
-give-same-old-excuses-supporting-obama-2012.

40. Glen Ford, "What Obama Has Wrought," *Black Agenda Report*, September 5, 2012, http://blackagendareport.com/content/what-obama-has
-wrought.

41. Glen Ford, "Angela Davis Has Lost Her Mind over Obama," *Black Agenda
Report*, March 27, 2012, http://blackagendareport.com/content/angela
-davis-lost-her-mind-over-obama.

42. Paul Street, "Obama Ticket Prices and the Invisible Ruling Class," *Black
Agenda Report*, March 11, 2014, http://www.blackagendareport.com
/content/obama-ticket-prices-and-invisible-ruling-class.

43. Thomas Frank, "Cornel West: 'He Posed as a Progressive and Turned
Out to Be Counterfeit. We Ended Up with a Wall Street Presidency,
a Drone Presidency," *Salon*, August 24, 2014, http://www.salon.com
/2014/08/24/cornel_west_he_posed_as_a_progressive_and_turned
_out_to_be_counterfeit_we_ended_up_with_a_wall_street
_presidency_a_drone_presidency/.

44. See Kimberly Nordyke, "Michael Moore Calls Obama's First Term
'Heartbreaking,' a 'Disappointment,'" *Hollywood Reporter*, October 25,
2011, http://www.hollywoodreporter.com/news/michael-moore-calls
-obamas-first-253253, and "Michael Moore's Harsh Prediction of President Obama's Legacy," *Atlanta Journal-Constitution*, September 10, 2014,
http://www.ajc.com/news/entertainment/michael-moores-harsh
-prediction-president-obamas-l/nhKHP/; Roger Hodge, *The Mendac-
ity of Hope: Barack Obama and the Betrayal of American Liberalism* (New
York: HarperCollins, 2010); and four articles by Diane McWhorter,
"Don't Punt on Torture," *USA Today*, February 11, 2009; "Redemption in Birmingham," *New York Times* (Sunday Review), July 9, 2011;
"Good and Evil in Birmingham," *New York Times*, January 20, 2013;
and "Obama's Atticus Finch Moment," *USA Today*, July 28, 2010. In
"Obama's Atticus Finch Moment," McWhorter writes: "So far, our first
black president has seemed inhibited rather than empowered by [our
racial] history. But only by transcending political necessity, risking fail-

ure for truth, will he earn a place alongside the heroes of our national mythology."

45. Chris Hedges, "The Obama Deception: Why Cornel West Went Ballistic," *Truthdig,* May 16, 2011, http://www.truthdig.com/report/item/the_obama_deception_why_cornel_west_went_ballistic_20110516; "Tavis Smiley, Cornel West on the 2012 Election and Why Calling Obama 'Progressive' Ignores His Record," *Democracy Now,* November 9, 2012, http://www.democracynow.org/2012/11/9/tavis_smiley_cornel _west_on_the.

46. "Cornel West: Al Sharpton 'the Bonafide House Negro of the Obama Plantation,'" *Real Clear Politics,* August 31, 2013, http://www.realclear politics.com/video/2013/08/31/cornel_west_al_sharpton_the_bona fide_house_negro_of_the_obama_plantation.html.

47. Lesley Stahl reported on the May 22, 2011, show, in a segment titled "Al Sharpton: The 'Refined' Agitator": "Sharpton told us that having a black president is a challenge: if he finds fault with Mr. Obama, he's aiding those who want to destroy him. So he has decided not to criticize the president about anything—even about black unemployment that's twice the national rate." When she asked him if he had told other black leaders not to criticize the president, Sharpton answered: "What I've told them is to be genuine about it. There are some blacks that said: 'He needs to go with a black agenda. He needs to do this.' He said when he was running he wasn't gonna do that. Duh. Surprise." When Stahl asked, despite Obama's not campaigning on the issue of black employment, why Sharpton wasn't proclaiming the need for more to be done, the civil rights leader said: "What I don't want to see is because he is black that we act like he's not the real president. 'He ought to be leading the black cause or the labor cause.' He's the president. To minimize who he is, I think, is an insult to the achievement of having him there." http://www.cbsnews.com/videos/al-sharpton-the-refined-agitator/.

48. See these claims, and my extensive engagement with West's criticisms of Obama and other black figures, in "The Ghost of Cornel West," *The New Republic,* May 2015, http://www.newrepublic.com/article/121550/cornel-wests-rise-fall-our-most-exciting-black-scholar-ghost.

49. Jonathan Alter, *The Center Holds: Obama and His Enemies* (New York: Simon & Schuster, 2013), p. 272.

50. Just a few months after Obama's first inauguration in 2009, I said to radio host Davey D—in terms that the left-wing *Black Agenda Report,*

my perennial critic, labeled "scathing words of criticism"—that we "are so grateful for having a black person in the office we don't demand anything of him," and "I expect the president of the United States to address issues of race." I argued that Obama has "fallen short and we must hold him accountable." In 2010, on a panel convened by talk show host Tavis Smiley, another vigorous Obama critic, I argued in Cornel West's presence that Obama is "Pharaoh, not Moses," and that black folk should not expect a politician to be a prophet even as we press him to respond to black needs. Later that year, on MSNBC, I said: "I think that we should push the president. This president runs from race like a black man runs from a cop. What we have to do is ask Mr. Obama to stand up and use his bully pulpit to help us. He is loath to speak about race." At a 2011 Congressional Black Caucus press conference held to criticize the Obama administration's failure to address chronic black unemployment, I said that this "is an American crisis that demands an American response at the highest echelons of our government, and that does include the White House. As gay and Latino and other Americans have done, we have to leverage our political power and voices to make this happen." In 2012, when I replaced Smiley as keynote speaker for a Martin Luther King Jr. luncheon for the city of Peoria, Illinois—Smiley was ousted because of his relentless and heated criticism of Obama—I began my speech by saying: "Tavis Smiley is a very dear friend of mine. I think he's an extraordinary human being . . . who's doing what he thinks is best . . . Dr. King would have taken some controversial stances, and did. He got disinvited too, trust me; Tavis is in good company. I support President Obama, but not without criticism, as you shouldn't. Nobody who's worth your support can be exempt from your critique." And in 2014, on *Face the Nation,* I argued that Obama should be far more vocal about the fires of unrest in Ferguson, Missouri, in the wake of the police killing of unarmed black youth Michael Brown: "This president knows better than most what happens in poor communities that have been antagonized, historically, by the hostile relationship between black people and the police department." I said that it "is not enough for him to come on national television and pretend that there's a false equivalency between police people who are armed, and black people [who] are vulnerable . . . He needs to use his bully pulpit to step up and articulate this as a vision." Following my remarks on television, I penned an op-ed in the *Washington Post,* where I claimed that Obama's remarks on Ferguson were largely tone-deaf and that he should provide more viable leadership on race and policing. These comments riled the White House and caused a heated exchange with a senior presidential adviser.

51. Ta-Nehisi Coates, "Color-Blind Policy, Color-Conscious Morality," *The Atlantic,* May 13, 2015, http://www.theatlantic.com/politics/archive /2015/05/color-blind-policy-color-conscious-morality/393227/. Coates has also argued that Obama is hamstrung by a set of facts about race that he can't afford to state: "What clearly cannot be said is that the events of Ferguson do not begin with Michael Brown lying dead in the street, but with policies set forth by government at every level. What clearly cannot be said is that the people of Ferguson are regularly plundered, as their grandparents were plundered, and generally regarded as a slush-fund for the government that has pledged to protect them. What clearly cannot be said is [that] the idea of superhuman black men who 'bulk up' to run through bullets is not an invention of Darren Wilson, but a staple of American racism. What clearly cannot be said is that American society's affection for nonviolence is notional. What cannot be said is that American society's admiration for Martin Luther King Jr. increases with distance, that the movement he led was bugged, smeared, harassed, and attacked by the same country that now celebrates him . . . What clearly cannot be said is that violence and nonviolence are tools, and that violence—like nonviolence—sometimes works . . . What cannot be said is that America does not really believe in nonviolence—Barack Obama has said as much—so much as it believes in order. What cannot be said is that there are very convincing reasons for black people in Ferguson to be nonviolent. But those reasons emanate from an intelligent fear of the law, not a benevolent respect for the law. The fact is that when the president came to the podium on Monday night there actually was very little he could say." Ta-Nehisi Coates, "Barack Obama, Ferguson, and the Evidence of Things Unsaid," *The Atlantic,* November 26, 2014, http://www .theatlantic.com/politics/archive/2014/11/barack-obama-ferguson -and-the-evidence-of-things-unsaid/383212/.

52. Coates wrote in "On the Death of Dreams," "If we are honest with ourselves we will see a president who believes in particular black morality, but eschews particular black policy." Ta-Nehisi Coates, "On the Death of Dreams," *The Atlantic,* August 29, 2013, http://www .theatlantic.com/politics/archive/2013/08/on-the-death-of-dreams /279157/.

53. Coates, "Color-Blind Policy, Color-Conscious Morality."

54. Ibid.

55. Ta-Nehisi Coates, "How the Obama Administration Talks to Black America," *The Atlantic*, May 20, 2013, http://www.theatlantic.com/politics/archive/2013/05/how-the-obama-administration-talks-to-black-america/276015/.

56. Jelani Cobb, *The Substance of Hope: Barack Obama and the Paradox of Progress* (New York: Walker Publishing, 2010).

57. Jelani Cobb, "Selma and Ferguson," *The New Yorker*, March 8, 2015, http://www.newyorker.com/news/news-desk/selma-and-ferguson.

58. Jelani Cobb, "A President and a King," *The New Yorker*, January 26, 2015, http://www.newyorker.com/magazine/2015/01/26/president-king.

59. Jelani Cobb, "Chronicle of a Riot Foretold," *The New Yorker*, November 25, 2014, http://www.newyorker.com/news/daily-comment/chronicle-ferguson-riot-michael-brown.

60. Jelani Cobb, "Requiem for a Dream," *The New Yorker*, August 28, 2013, http://www.newyorker.com/news/news-desk/requiem-for-a-dream.

61. Mary Frances Berry makes this claim in DeWayne Wickham, *Bill Clinton and Black America* (New York: Ballantine Publishing Group, 2012), p. 110: "I remember the dinner we had in the White House when we were discussing the 'mend it, don't end it' speech the president was planning to give on affirmative action. We were all sitting around the table talking about what he might say. Leon Higginbotham was there that night. So was Cornell [*sic*] West . . . [W]hen Clinton finally gave the speech at the National Archives, we were all invited there to hear him deliver it. His 'mend it, don't end it' policy was absolutely wonderful, given the way the courts had been cutting back on affirmative action, especially in the contracting area and higher education. For the Clinton administration to be able to go forward—not as much as it would have wished—but for him to find a way to continue to implement affirmative action was extraordinary." West's deep and detailed involvement in Clinton's policy of affirmative action, and the presidential speech to defend it, is exemplary of the very sort of principled participation that he now decries for other black figures involved with President Obama.

62. Cornel West with David Ritz, *Brother West: Living and Loving Out Loud, A Memoir* (New York: Smiley Books, 2009), p. 193.

2. "Invisible Man Got the Whole World Watching": Race, Bi-Race, Post-Race in the Obama Presidency

1. I would also endorse Obama in an article a few months later when *The Nation* magazine asked eight figures to support their chosen Democratic candidate among the eight politicians then running for president in a November 2007 issue nearly a year before the 2008 election. (Ellen Chesler, for instance, endorsed Hillary Clinton; Katherine S. Newman supported John Edwards; and Gore Vidal endorsed Dennis Kucinich.) See Michael Eric Dyson, "Barack Obama: A Visionary Candidate for a New America," *The Nation*, November 26, 2007.

2. Larry Blumenfeld, "Barack Obama in New Orleans," *Salon*, July 6, 2007, http://www.salon.com/2007/07/06/obama_172/.

3. Ibid.

4. Barack Obama's Democratic National Convention Keynote Address, Boston, July 27, 2004, *Washington Post*, http://www.washingtonpost.com/wp-dyn/articles/A19751-2004Jul27.html.

5. Rachel L. Swarns, "So Far, Obama Can't Take Black Vote for Granted," *New York Times*, February 2, 2007, http://www.nytimes.com/2007/02/11/opinion/11sun3.html?_r=0. Also see Brent Staples, "Decoding the Debate over the Blackness of Barack Obama," February 11, 2007, http://www.nytimes.com/2007/02/11/opinion/11sun3.html?_r=0; "The Obama Card: The Discussion of Race and the Senator's Candidacy Is Really About Whose Side He's On," *Los Angeles Times*, February 13, 2007, http://articles.latimes.com/2007/feb/13/opinion/ed-obama13; and Gary Younge, "Is Obama Black Enough?," *The Guardian*, March 1, 2007, http://www.theguardian.com/world/2007/mar/01/usa.uselections2008.

6. Stanley Crouch, "What Obama Isn't: Black Like Me," *New York Daily News*, November 2, 2006, http://www.nydailynews.com/archives/opinions/obama-isn-black-race-article-1.585922.

7. Debra J. Dickerson, "Colorblind: Barack Obama Would Be the Great Black Hope in the New Presidential Race — If He Were Actually Black," *Salon*, January 22, 2007, http://www.salon.com/2007/01/22/obama_161/.

8. For a brilliant rebuttal of Dickerson's (and, by extension, Crouch's) position, and one that takes into account a global conception of blackness that accentuates the complicated convergence of multiple ethnicities within black identity, see Joan Morgan, "Black Like Barack," in Sharpley-Whiting, *The Speech,* pp. 55–68. Morgan makes the point that Obama's "presidential run forced all Americans to grapple with the fact that 'black' in America is a diverse, multiethnic, sometimes biracial, and often bicultural experience that can no longer be confined to the rich but limited prism of U.S. slavery and its historical aftermath. As a first-generation black immigrant, I also know that Obama's precarious footing was caused . . . by the confusion and distrust this identity tends to provoke among whites and African Americans alike—precisely because it complicates, quite beautifully, not only existing constructs of race but all the traditional expectations, stereotypes, and explanations we have come to expect from discussions around what it means to be black in America" (pp. 59–60).

9. Victoria Brown, "In Solidarity: When Caribbean Immigrants Become Black," NBC News, March 2, 2015, http://www.nbcnews.com/news/nbcblk/solidarity-when-caribbean-immigrants-become-black-n308686; Jonathan Kaufman, "Help Wanted No Blacks Need Apply," *The Social Contract* 5, no. 4 (Summer 1995), http://www.thesocialcontract.com/artman2/publish/tsc0504/article_465_printer.shtml; Stephen Steinberg, "Immigration, African Americans, and Race Discourse," *New Politics* X-3 (Summer 2005), http://newpol.org/content/immigration-african-americans-and-race-discourse. Also see Joleen Kirschenman and Kathryn Neckerman, "'We'd Love to Hire Them but . . .': The Meaning of Race for Employers," in *The Urban Underclass,* ed. Christopher Jencks and Paul E. Peterson (Washington, D.C.: Brookings Institution, 1991); and Marcy C. Waters, *Black Identities: West Indian Dreams and American Realities* (Cambridge: Harvard University Press, 1999).

10. Morgan, "Black Like Barack." Even as she argues for broadening the palette of identities from which we paint black identity, Morgan acknowledges the thorny intraracial differences and political disputes between native-born blacks and immigrant blacks. "As first- and second-generation immigrants, we are often more conservative in our political ideology, are less likely to publicly embrace social programs like welfare, and tend to be very stalwart in our opinions about black complicity in our

own conditions. Racism for us is an undeniable reality, but it is also not the ultimate determinant. At our very core, we view America as a land of infinite possibilities because we know firsthand that it is possible to arrive in this country with nothing and build a life infinitely richer than the ones we left behind. We are, in short, a very up-from-the-bootstraps kind of people, a bit more Republican (although we tend not to vote that way), if not moderately Democratic, in nature than black political leaders care to recognize" (p. 64).

11. Obama, "Selma Voting Rights March Commemoration Speech."

12. Barack Obama, *Dreams from My Father: A Story of Race and Inheritance* (1995; repr., New York: Crown, 2004).

13. Carol B. Stack, *All Our Kin: Strategies for Survival in a Black Community* (1974; repr., New York: Basic Books, 1983).

14. James Baldwin, *The Fire Next Time* (1963; repr., New York: Vintage, 1993), p. 4. Writing to his nephew about the youth's grandfather—Baldwin's father—Baldwin states: "Well, he is dead, he never saw you, and he had a terrible life; he was defeated long before he died because, at the bottom of his heart, he really believed what white people said about him . . . You can only be destroyed by believing that you really are what the white world calls a *nigger*."

15. Obama, *Dreams*, p. 99.

16. Ibid., p. 87.

17. David Remnick, "The Joshua Generation: Race and the Campaign of Barack Obama," *The New Yorker*, November 17, 2008, http://www.new yorker.com/magazine/2008/11/17/the-joshua-generation. Also see David Remnick, *The Bridge: The Life and Rise of Barack Obama* (New York: Alfred A. Knopf, 2010), p. 4.

18. Frank Newport, "Obama Retains Strength Among Highly Educated," Gallup Poll, July 30, 2008, http://www.gallup.com/poll/109156/obama -retains-strength-among-highly-educated.aspx. Also see Janel Davis, "Is Education Level Tied to Voting Tendencies?," *PolitiFact*, November 5, 2012, http://www.politifact.com/georgia/statements/2012/nov/05 /larry-sabato/education-level-tied-voting-tendencies/. ("Based on the 2008 exit polls of Georgia, Virginia . . . and nationally, whites with a college degree supported Barack Obama at a higher rate than whites without a college degree.")

19. The most persuasive, and sophisticated, argument for Obama's being

"America's most progressive president since FDR," and that "electing a more compelling human being to the White House is probably impossible" in this nation, is made by Gary Dorrien in *The Obama Question: A Progressive Perspective* (Lanham, Md.: Rowman & Littlefield, 2012), p. 12.

20. Cited in Tim Wise, *Between Barack and a Hard Place: Racism and White Denial in the Age of Obama* (San Francisco: City Lights Publishers, 2009), p. 26.

21. "CNN's Candy Crowley Interviews President Barack Obama," CNN, December 21, 2014, http://cnnpressroom.blogs.cnn.com/2014/12/21/cnns-candy-crowley-interviews-president-barack-obama/.

22. Derrick Bell, *Faces at the Bottom of the Well: The Permanence of Racism* (New York: Basic Books, 1992).

23. Frank Rich, "In Conversation: Chris Rock," *New York,* November 30, 2014, http://www.vulture.com/2014/11/chris-rock-frank-rich-in-conversation.html.

24. For a powerful, empirically grounded argument about the effects of contemporary racial inequality—in an era when many racist barriers have fallen but racial inequality persists, not primarily because of the harmful things whites do to blacks, and other minorities, but because of the helpful things whites do for one another—see Nancy DiTomaso, *The American Non-Dilemma: Racial Inequality Without Racism* (New York: Russell Sage Foundation, 2013).

25. Helene Cooper, "Attorney General Chided for Language on Race," *New York Times,* March 7, 2009, http://www.nytimes.com/2009/03/08/us/politics/08race.html?_r=0.

26. Barack Obama, *The Audacity of Hope: Thoughts on Reclaiming the American Dream* (New York: Random House, 2006), pp. 363–364.

27. "Andrew Young Says Obama Lacks Experience to Be President, Bill Clinton 'As Black as Barack,'" Fox News, December 10, 2007, http://www.foxnews.com/story/2007/12/10/andrew-young-says-obama-lacks-experience-to-be-president-bill-clinton-as-black.html.

28. See Yolanda Putman, "Video: Chattanooga Pastor Challenges City to Deal with Violence at Martin Luther King Jr. Day March," *Chattanooga Times Free Press,* January 17, 2012, http://www.timesfreepress.com/news/news/story/2012/jan/17/martin-luther-king-day-challenge-chattanooga/68411/. In a sidebar timeline of King's life, there is this item:

"1953: Interviews to become minister at First Baptist Church on East Eighth Street in Chattanooga. Church overseers were concerned that, at age 24, he didn't have enough experience." Also see Lynda Edwards, "Chattanooga's Black History Sites Are Slowly Disappearing or Forgotten," *Chattanooga Times Free Press*, February 9, 2015, which states: "First Baptist Church: Martin Luther King Jr. interviewed for a job as minister of this church at 506 E. Eighth St., but the church thought he was too young at age 24." http://www.timesfreepress.com/news/life/entertainment/story/2015/feb/09/vanishing-history/287003/.

29. For an appreciation of how Jackson's progressive, big-tent, multiracial vision of the Democratic Party has prevailed, despite being for a time displaced by the neoliberalism of centrist Democrats like Bill Clinton, see Sam Tanenhaus, "Jesse Jackson Created the Modern Democratic Party," *Bloomberg View*, August 27, 2015, http://www.bloombergview.com/articles/2015-08-26/jesse-jackson-created-the-modern-democratic-party.

30. Jesse Jackson Jr., "Jesse Jr. to Jesse Sr.: You're Wrong on Obama, Dad," *Chicago Sun Times*, December 3, 2007, http://www.suntimes.com/news/commentary/.

31. Roddie A. Burris, "Jackson Slams Obama for 'Acting White,'" *The State*, September 19, 2007, shared on *Politico*, http://www.politico.com/story/2007/09/jackson-slams-obama-for-acting-white-005902.

32. "Jesse Jackson Disparages Barack Obama: Caught on Tape," *Huffington Post*, July 24, 2008, http://www.huffingtonpost.com/2008/07/16/jesse-jackson-caught-on-m_n_111732.html.

33. Ashley Southall, "Jesse Jackson Jr. Gets 30 Months, and His Wife 12, to Be Served at Separate Times," *New York Times*, August 14, 2013.

34. "Quotes in Reaction to Sean Bell Trial Verdict," abclocal.com, http://abclocal.go.com/story?section=news/local&id=6103450; "Obama Takes Questions on Sean Bell, Clyburn and Wright," *Washington Post*, April 25, 2008, http://voices.washingtonpost.com/44/2008/04/obama-takes-questions-on-sean.html.

35. James Baldwin made the claim, in conversation with psychologist Kenneth Clark, that "most cities are engaged in . . . something called urban renewal, which means moving Negroes out: it means Negro removal." See Kenneth B. Clark, "A Conversation with James Baldwin," in *Conver-*

sations with James Baldwin, ed. Fred L. Standley and Louis H. Pratt (Jackson: University of Mississippi Press, 1989), p. 42. Also see a video clip of Baldwin's interview with Clark in which Baldwin makes the statement about urban renewal as Negro removal: https://www.youtube.com/watch?v=T8Abhj17kYU .

36. Not all critics who argued that Obama wasn't black, but was instead biracial, were victims of such desires. Some maintained that the very categories of race that trapped us in the past continue to hold us captive, and that the refusal to see Obama as our first biracial president is the surest sign of our failure: "We are racially sophisticated enough to elect a non-white president, and we are so racially backward that we insist on calling him black," wrote the brilliant cultural critic Marie Arana in a thoughtful essay, "He's Not Black," in the *Washington Post.* "Progress has outpaced vocabulary." Arana pointed to her experience as a mixed-race Hispanic — white American and Peruvian — to argue for a more cosmopolitan view of race as glimpsed in the experience of Hispanic Americans. "Perhaps because we've been in this hemisphere two centuries longer than our northern brethren, we've had more time to mix it up. We are the product of el gran mestizaje, a wholesale cross-pollination that has been blending brown, white, black and yellow for 500 years — since Columbus set foot in the new world." One might conclude that Arana is right to criticize the refusal to embrace Obama's biracial heritage, and instead, lazily and retrogressively, call him black — a term, by the way, which Arana acknowledges that Obama embraces. What Arana may be overlooking in her critique is the politics of race that offers heightened esteem and greater privilege to the whiteness that is contained in such racial mixtures; biracialism becomes an appealing trait because it lessens blackness, generating an eagerness to embrace that whiteness, and therefore garnering greater acceptance in our culture while spurning the virtue of blackness and other nonwhite categories. Arana's Hispanic point of reference comes with built-in advantages in the dominant culture in regard to whiteness: there is in our culture's racial vocabulary the category of non-Hispanic white and, though far less frequently, non-Hispanic black, but none for non–African American white, or non-black white. See Marie Arana, "He's Not Black," *Washington Post,* November 30, 2008.

3. Black Presidency, Black Rhetoric:
Pharaoh and Moses Speak

1. "Who Sings It Better: Al Green or Obama?," *Today,* January 20, 2012, http://www.today.com/id/46069802/ns/today-today_news/t/who -sings-it-better-al-green-or-obama/#.Vem-xOvdVJo. The site also reports that Al Green later said of Obama's rendition that he "nailed it" and was "thrilled that the president even mentioned my name." Also see "Video: President Obama Sings Al Green's 'Let's Stay Together' at NYC Fundraiser," *US Weekly,* January 20, 2012, http://www.usmagazine .com/entertainment/news/president-obama-croons-al-greens-lets -stay-together-at-nyc-fundraiser-2012201. During a 2013 White House concert that celebrated Memphis soul, President Obama joked that he was one of the nation's premier impersonators of the soul legend. "Tonight, I am speaking not just as President, but as one of America's best-known Al Green impersonators," Obama said as the crowd laughed. http://www.washingtonexaminer.com/obama-im-one-of-americas -best-known-al-green-impersonators/article/2526759.

2. "Transcript of Obama's Remarks at the White House Correspondents Dinner," *Wall Street Journal,* April 29, 2012, http://blogs.wsj.com/wash wire/2012/04/29/transcript-of-obamas-remarks-at-the-white-house -correspondents-dinner/.

3. Aristotle, *On Rhetoric: A Theory of Civic Discourse,* trans. George A. Kennedy, 2nd ed. (New York: Oxford University Press, 2006); Kenneth Burke, *A Rhetoric of Motives* (Berkeley: University of California Press, 1969); Jeffrey K. Tullis, *The Rhetorical Presidency* (Princeton: Princeton University Press, 1988); Martin J. Medhurst, ed., *Beyond the Rhetorical Presidency* (College Station: Texas A&M University Press, 2004); Thomas W. Benson, *Writing JFK: Presidential Rhetoric and the Press in the Bay of Pigs Crisis* (College Station: Texas A&M University Press, 2003); *Bill Clinton on Stump, State, and Stage: The Rhetorical Road to the White House,* ed. Stephen A. Smith (Fayetteville: University of Arkansas Press, 1994); John M. Murphy, "Cunning, Rhetoric, and the Presidency of William Jefferson Clinton," in *The Presidency and Rhetorical Leadership,* ed. Leroy G. Dorsey (College Station: Texas A&M University Press, 2008), pp. 231–251; John Wilson, *Talking with the President: The Pragmatics of*

Presidential Language (New York: Oxford University Press, 2105); Justin S. Vaughn and Jennifer R. Mercieca, eds., *The Rhetoric of Heroic Expectations: Establishing the Obama Presidency* (College Station: Texas A&M University Press, 2014).

4. Elaine B. Richardson and Ronald L. Jackson II, eds., *African American Rhetoric(s): Interdisciplinary Perspectives* (Carbondale: Southern Illinois University Press, 2007); Keith Gilyard, *True to the Language Game: African American Discourse, Cultural Politics, and Pedagogy* (New York: Routledge, 2011).

5. Barack Obama in Sumter, S.C., YouTube, January 24, 2008, https://www.youtube.com/watch?v=gh69Zi2rV-U; also Ben Smith, "An Unlikely Echo," *Politico*, January 27, 2008, http://www.politico.com/blogs/bensmith/0108/An_unlikely_echo.html.

6. Late into Obama's second term, as he rode a crest of political successes, the Obama-Reagan comparisons, occasionally prompted by Obama himself in private conversation, appear far less troublesome, and instead potentially place him within shouting distance of the sort of transformative presidencies he said in January 2008 that Richard Nixon and Bill Clinton lacked. See Linda Feldmann, "Is Obama the Democrats' Reagan?," *Christian Science Monitor*, September 6, 2105, http://www.csmonitor.com/USA/Politics/2015/0906/Is-Obama-the-Democrats-Reagan.

7. Excerpt from *Malcolm X* (1992), directed by Spike Lee, https://www.youtube.com/watch?v=DV7yx2y3TtY.

8. *New Yorker* editor David Remnick defended the cover artwork by Barry Blitt as satire: "Our cover 'The Politics of Fear' combines a number of fantastical images about the Obamas and shows them for the obvious distortions they are. The burning flag, the nationalist-radical and Islamic outfits, the fist-bump, the portrait on the wall—all of them echo one attack or another. Satire is part of what we do, and it is meant to bring things out into the open, to hold up a mirror to prejudice, the hateful, and the absurd. And that's the spirit of this cover." See Tobin Harshaw, "Obama's Cover Flap," *New York Times*, July 14, 2008, http://opinionator.blogs.nytimes.com/2008/07/14/obamas-cover-flap/comment-page-5/. Blitt briefly responded as well: "I think the idea that the Obamas are branded as unpatriotic [let alone as terrorists] in certain sectors is preposterous. It seemed to me that depicting the

concept would show it as the fear-mongering ridiculousness that it is."
See Nico Pitney, "Barry Blitt Defends His New Yorker Cover Art of
Obama," *Huffington Post*, July 21, 2008, http://www.huffingtonpost.com
/2008/07/13/barry-blitt-addresses-his_n_112432.html.

9. "Conservative Outrage Builds over Obama's 'Race-Baiting' Comments
on Shooting of Unarmed Black Teen After President Obama Said, 'If I
Had a Son, He'd Look Like Trayvon,'" *Daily Mail*, March 23, 2012, http://
www.dailymail.co.uk/news/article-2119340/Trayvon-Martin-case
-Newt-Gingrich-slams-Obamas-disgraceful-comments-shooting
.html. Also see David Weigel, "'If Obama Had a Son He Would Look
Like Aaron Alexis,'" *Slate*, September 16, 2013, http://www.slate.com
/blogs/weigel/2013/09/16/navy_yard_shooting_suspect_aaron_alexis
_is_black_so_obviously_twitter_jerks.html.

10. Obama, Democratic National Convention Keynote Address.

11. Barack Obama's Iowa Caucus Speech, *New York Times*, January 3, 2008,
http://www.nytimes.com/2008/01/03/us/politics/03obama-tran
script.html?pagewanted=all.

12. Barack Obama's New Hampshire Primary Speech, *New York Times*, January 8, 2008, http://www.nytimes.com/2008/01/08/us/politics/08text
-obama.html?pagewanted=all.

13. At the time of Biden's comment, I told the *New York Times* that "historically, [the word "articulate"] was meant to signal the exceptional Negro.
The implication is that most black people do not have the capacity to
engage in articulate speech, when white people are automatically assumed to be articulate." See Lynette Clemetson, "The Racial Politics of
Speaking Well," *New York Times*, February 4, 2007, http://www.nytimes
.com/2007/02/04/weekinreview/04clemetson.html?gwt=pay.

14. "Biden's Description of Obama Draws Scrutiny," CNN, February 9,
2007, http://www.cnn.com/2007/POLITICS/01/31/biden.obama/.

15. H. Samy Alim and Geneva Smitherman, *Articulate While Black: Barack
Obama, Language, and Race in the U.S.* (New York: Oxford University
Press, 2012).

16. Philip Elliott, "Harry Reid 'Negro' Comment: Reid Apologizes for 'No
Negro Dialect' Comment," *Huffington Post*, March, 18, 2010, http://www
.huffingtonpost.com/2010/01/09/harry-reid-negro-comment-_n
_417406.html.

17. Dr. Frederick G. Sampson was noted by *Ebony* magazine as one of the

fifteen greatest black preachers in America in 1984 and again in 1993. *Ebony* cited Sampson for his "depth of exegetical insight, brilliance of illustrations and captivating style of communication," adding that Sampson "laces his sermons with moving, real-life illustrations and is highly dramatic with respect to both language and gestures." See "The 15 Greatest Black Preachers," *Ebony*, November 1993, p. 168.

18. Wright, according to *Ebony*, represents "the first generation of African-American preachers who blend a Pentecostal flavor with social concerns in their pulpit discourse." Wright "gives a contemporary, African-American, Afrocentric flavor to the traditional Black shout." A Wright sermon "is a four-course meal: spiritual, biblical, cultural, prophetic." Ibid., p. 157.

19. David Garrow, *Bearing the Cross: Martin Luther King, Jr., and the Southern Christian Leadership Conference* (1986; repr., New York: Perennial Classics, 2004), p. 622.

20. I made this point on *Meet the Press* in 2008 in a discussion of King's legacy with the host, the late, great Tim Russert, and fellow guests, former NBC News anchor Tom Brokaw and Ambassador Andrew Young. I spoke about how King's rhetoric in the black church was dramatically different from — and far more radical than — his messages for white America: "When you heard Jeremiah Wright, what you heard was the latter-day Martin Luther King Jr. When you hear Barack Obama, you hear Dr. King up to 1965. In black churches, Martin Luther King Jr. said, 'We have been subject to American genocide.' He also went on to say that he didn't want to be treated the same way the Japanese brothers and sisters [were] when they were put in the concentration camps. And the sermon he was going to deliver, Tim, the next Sunday, were he to live, found in the effects after he was murdered, was a sermon called 'Why America May Go to Hell.' That's the Martin Luther King Jr. with which the broad swath of America is not familiar, and they don't understand, within the black church, the articulation of a theological tradition that responds to hatred, doesn't respond in hate but prophetic anger and then, ultimately, love — love enough to speak justice to the nation. Justice is what love sounds like when it speaks in public, and Martin Luther King Jr. did this when he talked specifically to black churches." "MLK's Impact on the World," *Meet the Press*, April 6, 2008, http://www.nbcnews.com/video/meet-the-press/23981403#23981403.

21. Michael Eric Dyson, *I May Not Get There with You: The True Martin Luther King, Jr.* (2000; repr., New York: Touchstone, 2001), pp. 87–88.

22. James M. Washington, ed., *A Testament of Hope: The Essential Writings and Speeches of Martin Luther King, Jr.* (1986; repr., New York: HarperOne, 2003), pp. 264–265.

23. Dyson, *I May Not*, p. 40.

24. Ibid., pp. 38–39.

25. Richard Lischer, *The Preacher King: Martin Luther King Jr. and the Word That Moved America* (New York: Oxford University Press, 1997), p. 159.

26. Ibid., p. 158.

27. Brian Ross and Rehab El-Buri, "Obama's Pastor: God Damn America, U.S. to Blame for 9/11," ABC News, March 13, 2008, http://abcnews .go.com/Blotter/DemocraticDebate/story?id=4443788. Also see Kumarini Silva, "Browning Our Way to Post-Race: Identity, Identification, Securitization of Brown," in *American Identity in the Age of Obama*, ed. Amilcar Antonio Barreto and Richard L. O'Bryant (New York: Routledge, 2014), p. 140.

28. Obama, *Dreams*, p. 294.

29. For a brilliant reading of the broader context to which Obama was hardly able to allude, see Obery M. Hendricks Jr., "A More Perfect (High-Tech) Lynching: Obama, the Press, and Jeremiah Wright," in Sharpley-Whiting, *The Speech*, pp. 155–183.

30. "Barack Obama Interview on March 16, 2008," *Chicago Tribune*, March 16, 2008, http://www.chicagotribune.com/news/chi-obamafullwebmar 16-archive-story.html#page=9.

31. For a convincing argument about how Martin Luther King Jr. channeled in his oratory the subversive meanings of the dream metaphor expressed in the poetry of Langston Hughes, see W. Jason Miller, *Origins of the Dream: Hughes's Poetry and King's Rhetoric* (Gainesville: University Press of Florida, 2015).

32. Obama, "A More Perfect Union," pp. 237–251.

33. Katharine Q. Seelye and Julie Bosman, "Ferraro's Obama Remarks Become Talk of Campaign," *New York Times*, March 12, 2008, http://www .nytimes.com/2008/03/12/us/politics/12campaign.html.

34. Ben Smith, "A Ferraro Flashback," *Politico*, March 11, 2008, http://www .politico.com/blogs/ben-smith/2008/03/a-ferraro-flashback-006934.

35. See "Malcolm X at Harvard University," March 18, 1964, http://malcolm

xfiles.blogspot.com/2013/06/malcolm-x-at-harvard-university-march
.html. For a sophisticated psycho-biographical study of Malcolm X, see
Eugene Victor Wolfenstein, *The Victims of Democracy: Malcolm X and the
Black Revolution* (1981; repr., New York: Guilford Press, 1993).

36. Jeremiah Wright, "The Day of Jerusalem's Fall" (partial transcript), *The
Guardian,* March 27, 2008, http://www.theguardian.com/commentis
free/2008/mar/27/thedayofjerusalemsfall.

37. James H. Jones, *Bad Blood: The Tuskegee Syphilis Experiment,* rev. ed. (New
York: Free Press, 1993).

38. Trymaine Lee, "Tavis Smiley: 'I Don't Get Intimidated by Haters,'"
Huffington Post, August 12, 2011, http://www.huffingtonpost.com/2011
/08/12/tavis-smiley-i-dont-get-i_n_925920.html.

39. For my take on the black prophetic tradition, see my article "The Ghost
of Cornel West," *The New Republic,* May 2015, http://www.newrepublic
.com/article/121550/cornel-wests-rise-fall-our-most-exciting-black
-scholar-ghost.

40. For the single best essay I've read on the frustration over the lack of
positive outcomes from black faces in high political places, see Keeanga-
Yamahtta Taylor, "In Baltimore and Across the Country, Black Faces
in High Places Haven't Helped Average Black People," *In These Times,*
April 29, 2015, http://inthesetimes.com/article/17888/baltimore_riots
_black_politicians.

41. "Obama Interview on March 16, 2008."

42. Wright seems to have understood the principle of forgiveness knowledge when he recalled, in a question after his speech before the National
Press Club on April 28, 2008: "Several of my white friends and several
of my white, Jewish friends have written me and said to me. They've
said, 'You're a Christian. You understand forgiveness. We both know
that, if Senator Obama did not say what he said, he would never get
elected.'" Also, when asked how he felt about Obama's distancing himself from him, Wright responded: "He didn't distance himself. He *had* to
distance himself, because he's a politician, from what the media was saying I had said, which was anti-American." "Reverend Wright at the National Press Club, April 28, 2008," *New York Times,* http://www.nytimes
.com/2008/04/28/us/politics/28text-wright.html?pagewanted=all.

43. Thomas Beaumont, "Up-Close Obama Urges Compassion in Mideast:
He Backs Loosening of Restrictions on Palestinian Aid," *Des Moines Reg-*

ister, March 12, 2007, http://www.factcheck.org/2007/04/democratic
-candidates-debate/. Later, at the first full Democratic presidential
debate of the 2008 campaign on the campus of South Carolina State
University in Orangeburg on April 26, 2007, moderator Brian Williams
asked Obama if he stood by the remark. Obama replied: "Well, keep in
mind what the remark actually, if you had the whole thing, said. And
what I said is nobody has suffered more than the Palestinian people
from the failure of the Palestinian leadership to recognize Israel, to re-
nounce violence, and to get serious about negotiating peace and security for the region." The *Des Moines Register,* however, quotes Obama as
attributing Palestinian suffering to "the stalled peace efforts with Israel"
and not to lapses in Palestinian leadership. The report reads: "Obama
told the Muscatine-area party activists that he supports relaxing restrictions on aid to the Palestinian people. He said they have suffered the
most as a result of stalled peace efforts with Israel. 'Nobody is suffering
more than the Palestinian people,' Obama said while on the final leg of
his weekend trip to eastern Iowa."

44. For a brilliant discussion of the history, themes, disputes, arguments,
politics, and moral trajectories of black power and black self-determina-
tion — and the effort to improve black life with black hands — see Peniel
Joseph, *Waiting 'Til the Midnight Hour: A Narrative History of Black Power
in America* (2006; repr., New York: Holt Paperbacks, 2007).

45. "Transcript of the Keynote Address by Ann Richards, the Texas
Treasurer," *New York Times,* July 19, 1988, http://www.nytimes.com
/1988/07/19/us/transcript-of-the-keynote-address-by-ann-richards
-the-texas-treasurer.html.

46. Dyson, *I May Not,* p. 12.

47. "Reverend Wright at the National Press Club, April 28, 2008."

48. Transcript, *Bill Moyers Journal,* April 25, 2008, http://www.pbs.org
/moyers/journal/04252008/transcript1.html.

49. "Transcript of Jeremiah Wright's Speech to NAACP," CNN, April 28,
2008, http://www.cnn.com/2008/POLITICS/04/28/wright.tran
script/.

50. As theologian Paul Tillich argues, modern languages "have only one
word for 'time.' The Greeks had two words, *chronos* and *kairos. Chronos*
is clock time, time which is measured, as we have it in words like "chronology" and "chronometer." *Kairos* is not the quantitative time of the

clock, but the qualitative time of the occasion, the right time . . . There are things that happen when the right time, the *kairos*, has not yet come. *Kairos* is the time which indicates that something has happened which makes an action possible or impossible. We all experience moments in our lives when we feel that now is the right time to do something, now we are mature enough, now we can make the decision. This is the *kairos*." Paul Tillich, *A History of Christian Thought: From Its Judaic and Hellenistic Origins to Existentialism*, ed. Carl E. Braaten (New York: Simon & Schuster/Touchstone, 1972), p. 1.

51. Richard Lentz, *Symbols, the News Magazines, and Martin Luther King* (Baton Rouge: Louisiana State University Press, 1990), esp. pp. 176–177.

52. "Priest Apologizes for Mocking Clinton While at Obama Church," CNN, May 30, 2008, http://www.cnn.com/2008/POLITICS/05/30/obama.pfleger/.

53. Susan Taylor, "What We Can Do to Reclaim Black Children," *L.A. Watts Times*, February 14, 2013, http://www.lawattstimes.com/index.php?option=com_content&view=article&id=920:what-we-can-do-to-reclaim-black-children&catid=24&Itemid=119.

54. Luke 12:48 (New Revised Standard Version).

4. Re-Founding Father: Patriotism, Citizenship, and Obama's America

1. Allison Hoffmann, "Oprah Hosts Obama in Star-Studded Event," *Washington Post*, September 9, 2007, http://www.washingtonpost.com/wp-dyn/content/article/2007/09/08/AR2007090801197.html.

2. Chris Rock has told the story several times, perhaps most famously on the HBO special *The Black List* (vol. 1), 2009, https://www.youtube.com/watch?v=8q1XUo7BIOo.

3. Randall Kennedy draws a parallel between his father's bitter resistance to American patriotism and Jeremiah Wright's pungent racial views in *The Persistence of the Color Line: Racial Politics and the Obama Presidency* (New York: Pantheon Books, 2011), pp. 161–195.

4. Michael Eric Dyson, "Understanding Black Patriotism," *Time*, April 24, 2008, http://www.webcitation.org/5XKexgdfT.

5. "Frederick Douglass: 'The Meaning of July Fourth for the Negro,'" Zinn Education Project, https://zinnedproject.org/materials/frederick-douglass-the-meaning-of-july-fourth-for-the-negro/.

6. Langston Hughes, "Let America Be America Again," http://www.poets .org/poetsorg/poem/let-america-be-america-again.

7. Lentz, *Symbols*, p. 239; Dyson, *I May Not*, p. 62.

8. Jordan Goodman, *Paul Robeson: A Watched Man* (New York: Verso, 2013); David Garrow, *The FBI and Martin Luther King, Jr.: From "Solo" to Memphis* (New York: W. W. Norton, 1981).

9. Martin Luther King Jr., "I See the Promised Land," in *A Testament of Hope: The Essential Writings and Speeches of Martin Luther King, Jr.*, ed. James M. Washington (New York: HarperOne, 2003), p. 282.

10. James Baldwin, *Notes of a Native Son* (1955; repr., Boston: Beacon Press, 1984), p. 9.

11. When quizzed at the National Press Club about whether his sermons were unpatriotic, Wright responded: "I feel that those citizens who say that have never heard my sermons, nor do they know me. They are unfair accusations taken from sound bites and that which is looped over and over again on certain channels. I served six years in the military. Does that make me patriotic? How many years did [former Vice President Dick] Cheney serve?" "Reverend Wright at the National Press Club, April 28, 2008," *New York Times,* http://www.nytimes .com/2008/04/28/us/politics/28text-wright.html?pagewanted=all.

12. Two military veterans—Lawrence Korb, from the U.S. Navy, and Ian Moss, from the Marine Corps—penned an article in the *Chicago Tribune* to clarify and defend Jeremiah Wright's notable record of service in the military not long after it was integrated. See Lawrence Korb and Ian Moss, "Factor Military Duty into Criticism," *Chicago Tribune,* April 3, 2008, http://www.webcitation.org/5WoiVsN1X.

13. For more of my thoughts on the subject, see Michael Eric Dyson, *Pride: The Seven Deadly Sins* (New York: Oxford University Press, 2006), pp. 85–118.

14. "Michelle Obama Clarifies: I've Always Loved My Country," CNN, February 20, 2008, http://politicalticker.blogs.cnn.com/2008/02/20 /michelle-obama-clarifies-ive-always-loved-my-country/. Initially, what she said was, "For the first time in my adult life, I am really proud of my country, because it feels like hope is making a comeback . . . not just because Barack has done well, but because I think people are hungry for change." It seems a reasonable statement to make for a member of an oppressed minority group that has been denied critical opportunities

and systematically prevented from flourishing on a number of fronts. Now that the country had shown a profound willingness to change, Michelle Obama could understandably take pride in the country's ability to move forward, seen in the groundswell of support for her husband's historic candidacy. It seemed disingenuous for her critics to deny the legacy of inequality that might make a member of a minority group not as happy, or as proud, as she might otherwise be if that group enjoyed the full benefits and ripe fruit of democracy that the majority has had access to all along. Comedian Chris Rock captured the ambivalence that many blacks feel with regard to being a member of a society that has begun to open doors yet has a history of injustice: "If you're black, America's like the uncle that paid your way through college — but molested you." http://www.theguardian.com/film/2005/feb/26/award sandprizes.oscars2005.

15. Herb Galewitz, ed., *Patriotism: Quotations from Around the World* (Mineola, N.Y.: Dover, 2003), p. 27.

16. Salamishah Tillet, *Sites of Slavery: Citizenship and Racial Democracy in the Post–Civil Rights Imagination* (Durham: Duke University Press, 2012). Tillet argues that "civic estrangement occurs because [African Americans] have been marginalized or underrepresented in the civic myths, monuments, narratives, icons, creeds and images of the past that constitute, reproduce and promote an American national identity" (p. 18).

17. David Wright and Sunlen Miller, "Obama Dropped Flag Pin in War Statement," ABC News, October 4, 2007, http://abcnews.go.com/Poli tics/story?id=3690000.

18. David C. Anderson, *Crime and the Politics of Hysteria: How the Willie Horton Story Changed American Justice* (New York: Crown, 1995).

19. "McCain Distances Himself from Supporter's Comments," CNN, February 26, 2008, http://politicalticker.blogs.cnn.com/2008/02/26 /mccain-distances-himself-from-supporters-comments/.

20. Barack Obama, "Full Script of Obama's Speech," CNN, July 24, 2008, http://edition.cnn.com/2008/POLITICS/07/24/obama.words/.

21. Jonathan Weisman and Juliet Eilperin, "Race Moves to Center Stage," *Washington Post,* August 1, 2008, http://www.washingtonpost.com /wp-dyn/content/article/2008/08/01/AR2008080102970.html.

22. Barack Obama's Speech in Independence, Mo., *New York Times,* June 30, 2008, http://www.nytimes.com/2008/06/30/us/politics/30text -obama.html?pagewanted=all&_r=0.

23. Barack Obama, "What I See in Lincoln's Eyes," *Time,* July 4, 2005, http://content.time.com/time/magazine/article/0,9171,1077287,00.html.

24. Thanks to Noelle Braddock.

25. Dinesh D'Souza, *The Roots of Obama's Rage* (Washington, D.C.: Regnery, 2010), and *Obama's America: Unmaking the American Dream* (Washington, D.C.: Regnery, 2012).

26. Grace Wyler, "John Sununu: 'I Wish This President Would Learn How to Be an American,'" *The Atlantic,* July 17, 2012, http://www.theatlantic.com/politics/archive/2012/07/john-sununu-i-wish-this-president-would-learn-how-to-be-an-american/259948/.

27. Joy Lin, "Gingrich: Obama Is Most Dangerous President in American History," Fox News, February 20, 2012, http://www.foxnews.com/politics/2012/02/20/gingrich-obama-is-most-dangerous-president-in-american-history/.

28. Daniel Halper, "Senator: 'I Have Not Been Impressed with [Holder's] Intelligence,'" *Weekly Standard,* June 13, 2012, http://www.weeklystandard.com/blogs/senator-i-have-not-been-impressed-holders-intelligence_647204.html.

29. Kevin Robillard, "McCain: Rice 'Not Qualified' for State," *Politico,* November 14, 2012, http://www.politico.com/story/2012/11/mccain-rice-not-qualified-for-state-083824.

30. Jamelle Bouie, "From Good to Great," *Slate,* September 25, 2014, http://www.slate.com/articles/news_and_politics/politics/2014/09/eric_holder_resigning_as_attorney_general_his_justice_department_was_a_staunch.html.

31. Of the quickly growing body of literature on the modern-day Tea Party, here are some of the best, and the ones that have influenced my view of the Tea Party in these pages: Matthew W. Hughey and Gregory S. Parks, *The Wrongs of the Right: Language, Race, and the Republican Party in the Age of Obama* (New York: NYU Press, 2014); David Corn, *Showdown: The Inside Story of How Obama Fought Back Against Boehner, Cantor, and the Tea Party* (New York: William Morrow, 2012); Meghan A. Burke, *Race, Gender, and Class in the Tea Party: What the Movement Reflects About Mainstream Ideologies* (Lanham, Md.: Lexington Books, 2015); Kate Zernike, *Boiling Mad: Inside Tea Party America* (New York: Times Books, 2010); Anthony DiMaggio, *The Rise of the Tea Party: Political Discontent and Corporate Media in the Age of Obama* (New York: Monthly Review Press, 2011); and Jill Lepore, *The Whites of Their Eyes: The Tea Party's Revolu-*

tion and the Battle over American History (Princeton: Princeton University Press, 2010).

32. Haley Crum, "Obama Provides Birth Certificate — on a Button," *Washington Post,* August 27, 2012, http://www.washingtonpost.com/news/post-politics/wp/2012/08/27/obama-provides-birth-certificate-on-a-button/.

33. Jim Lobe, "As Iraq Anniversary Fades, America's 'Strategic Narcissism' Stands Out," Inter Press Service, March 24, 2013, http://www.common dreams.org/news/2013/03/24/iraq-anniversary-fades-americas-strate gic-narcissism-stands-out.

34. Adam Shah, "Having Attacked Obama for Overseas 'Apology Tour,' Conservative Media Now Attack Him for Not Going to Berlin," *Media Matters,* November 9, 2009, http://mediamatters.org/research/2009/11/09/having-attacked-obama-for-overseas-apology-tour/156748.

35. Barack Obama, "Remarks by President Obama at Strasbourg Town Hall," Strasbourg, France, https://www.whitehouse.gov/the-press-office/remarks-president-obama-strasbourg-town-hall.

36. Barack Obama, "Remarks by President Obama to the Turkish Parliament," Ankara, April 6, 2009, https://www.whitehouse.gov/the-press-office/remarks-president-obama-turkish-parliament.

37. Barack Obama, "Remarks by the President at Cairo University, 6-04-09," June 4, 2009, https://www.whitehouse.gov/the-press-office/remarks-president-cairo-university-6-04-09.

38. Macon Phillips, "Osama Bin Laden Dead," May 2, 2011, https://www.whitehouse.gov/blog/2011/05/02/osama-bin-laden-dead.

39. Barack Obama, "Statement by the President on ISIL," The White House, September 10, 2014, https://www.whitehouse.gov/the-press-office/2014/09/10/statement-president-isil-1.

40. Alex Spillius, "Barack Obama Tells Africa to Stop Blaming Colonialism for Problems," *Daily Telegraph* (London), July 9, 2009, http://www.telegraph.co.uk/news/worldnews/africaandindianocean/5778804/Barack-Obama-tells-Africa-to-stop-blaming-colonialism-for-problems.html.

41. Barack Obama, "Remarks by the President to the Ghanaian Parliament," Accra, July 11, 2009, https://www.whitehouse.gov/the-press-office/remarks-president-ghanaian-parliament.

42. "Interview of the President by AllAfrica.com, 7-2-09," https://www

.whitehouse.gov/the-press-office/interview-president-allafrica
com-7-2-09.

43. Obama, "Remarks to the Ghanaian Parliament."

44. Stephanie Hanson, "Angola's Political and Economic Development,"
Council on Foreign Relations, July 21, 2008, http://www.cfr.org/world
/angolas-political-economic-development/p16820.

45. Terra Lawson-Remer and Joshua Greenstein, "Beating the Resource
Curse in Africa: A Global Effort," Council on Foreign Relations, August
2012, http://www.cfr.org/africa-sub-saharan/beating-resource-curse
-africa-global-effort/p28780.

46. Nancy Birdsall and John Nellis, "Winners and Losers: Assessing the Dis-
tributional Impact of Privatization," World Development 31, no. 10 (October 2003): 1617–33.

47. Marcy C. Diang, "Colonialism, Neoliberalism, Education and Culture
in Cameroon," paper no. 52 (Fall 2013), DePaul University College of
Education, http://via.library.depaul.edu/soe_etd/52; David William
Pear, "Africa: Incredible Wealth, Exploitation, Corruption and Poverty
for Its People," Real News Network, January 27, 2014; http://therealnews
.com/t2/component/content/article/170-more-blog-posts-from
-david-william-pear/1944-africa-incredible-wealth-exploitation
-corruption-and-poverty-for-its-people-.

48. David Levering Lewis, The Race to Fashoda: European Colonialism and African Resistance in the Scramble for Africa (New York: Weidenfeld & Nicol-
son, 1987).

49. Annika McGinnis, "Obama Urges African Nations Not to Make Economic 'Excuses,'" Reuters, July 28, 2014, http://www.reuters.com/article
/2014/07/28/us-africa-obama-idUSKBN0FX1WW20140728.

50. For competing arguments about the use of Africom by the U.S. military,
see Robert Moeller, "The Truth About AFRICOM," Foreign Policy, July 21,
2010, http://foreignpolicy.com/2010/07/21/the-truth-about-africom/;
and Mark P. Fancher, "Dr. Che Guevera's Prescription for Africa's
AFRICOM Headache," Black Agenda Report, July 7, 2015, http://www
.blackagendareport.com/che_guevara_AFRICOM.

51. Bret Stephens, "Obama Gets It Right on Africa," Wall Street Journal, July
15, 2009, http://www.wsj.com/articles/SB124753013433935785.

52. Toni Morrison, "On the Backs of Blacks," Time, December 2, 1993,
http://content.time.com/time/magazine/article/0,9171,979736,00
.html.

53. W. E. B. Du Bois, *The Souls of Black Folk* (1903; repr., Mineola, N.Y.: Dover, 1994), pp. 2–3.

5. The Scold of Black Folk: The Bully Pulpit and Black Responsibility

1. Obama, *Audacity,* p. 387.

2. See Shirley Sherrod with Catherine Whitney, *The Courage to Hope: How I Stood Up to the Politics of Fear* (New York: Atria Books, 2012). Given the play in her book title on Obama's "audacity of hope," one wonders if Sherrod isn't suggesting that audaciousness isn't enough in a political atmosphere that demands the courage of one's convictions.

3. Tony Perry, "Shirley Sherrod Vows to Sue Conservative Blogger Who Misrepresented Her Remarks," *Los Angeles Times,* July 29, 2010, http://latimesblogs.latimes.com/lanow/2010/07/shirley-sherrod-speech.html.

4. Sheryl Gay Stolberg, "For Obama, Nuance on Race Invites Questions," *New York Times,* February 8, 2010, http://www.nytimes.com/2010/02/09/us/politics/09race.html.

5. "Racial Disparities in Disbursement of Stimulus Funding," Hispanic Ad.com, January 22, 2010, http://hispanicad.com/blog/news-article/had/government/racial-disparities-disbursement-stimulus-funding.

6. Mark Kantrowitz, "The Distribution of Grants and Scholarships by Race," FinAid.org, September 2, 2011, p. 8, http://www.finaid.org/scholarships/20110902racescholarships.pdf.

7. William Julius Wilson, "Race-Neutral Policies and the Democratic Coalition," *The American Prospect,* December 4, 2000, http://prospect.org/article/race-neutral-policies-and-democratic-coalition.

8. William Julius Wilson, *More Than Just Race: Being Black and Poor in the Inner City* (New York: W. W. Norton, 2009), pp. 141–142.

9. Stolberg, "For Obama, Nuance on Race Invites Questions."

10. As I argue elsewhere: "There were at least a few competing versions of the universal floating around. The trick was to incorporate one version of universalism, black rights, into the legal arc of another version of universalism, white privilege, while preserving the necessary illusion of neutrality on which such rights theoretically depended. Hence a philosophical principle — what the philosopher Hegel might call a 'concrete universal' — was transformed into a political strategy, allowing both

whites and blacks to preserve their specific stake in a universal value: democracy. To miss this process — that is, to mistake politics for philosophical principles, or, in turn, to disregard their symbiotic relationship in shaping American democracy — is to distort fatally the improvisational, ramshackle, halt-and-leap fashion by which American politics achieves its goals." Dyson, *I May Not*, p. 26.

11. Michael Eric Dyson, *I May Not Get There with You: The True Martin Luther King, Jr.* (New York: Free Press, 2000), p. 18.

12. Lauren Victoria Burke, "Is Black America Better Off Under Obama?" *Black Press USA,* January 5, 2015, http://www.blackpressusa.com/is -black-america-better-off-under-obama/#sthash.XlB2XRKv.dpbs; Zak Cheney-Rice, "13 Startling Numbers Reveal the Reality of Black America Under Obama," *IdentitiesMic,* December 19, 2014, http://mic.com /articles/106868/13-numbers-that-highlight-the-difference-between -obama-s-post-racial-dream-and-reality. Also see the predictable — and right-for-the-wrong-reasons — attack from the right: Deroy Murdock, "Black Americans Are Worse Off Under Obama," *National Review,* May 16, 2014, http://www.nationalreview.com/article/378087/black-americans -are-worse-under-obama-deroy-murdock; and Jennifer G. Hickey, "Race Gap: Blacks Fall Further Behind Under Obama," *Newsmax,* January 8, 2014, http://www.newsmax.com/Newsfront/obama-blacks-poverty -education/2014/01/08/id/545866/. And, of course, from the black left: Dr. Reginald Clark, "The Expansion of Black American Misery Under Barack Obama's Watch," *Black Agenda Report,* February 9, 2013, http:// www.blackagendareport.com/content/expansion-black-american -misery-under-barack-obama's-watch.

13. "Transcript of President Obama's Remarks at Year-End Press Conference (All Women Questioners Edition)," *Daily Kos,* December 19, 2014, http://www.dailykos.com/story/2014/12/19/1353032/-Transcript-of -President-Obama-s-remarks-at-year-end-press-conference-all-women -questioners-edition#.

14. Ruth Simon and Tom McGinty, "Loan Rebound Misses Black Businesses," *Wall Street Journal,* March 14, 2014, http://www.wsj.com/articles /SB10001424052702304585004579417021571596610.

15. When Obama came into office in 2009, the national unemployment rate was 7.8 percent, and by April 2014 it had fallen to 6.3 percent, a marked improvement. The black unemployment rate dropped far less

markedly, going from 12.7 percent to 11.6 percent, nearly double the national average. For blacks between the ages of sixteen and nineteen, the unemployment rate grew from 35.3 percent to 36.8 percent. To be fair, blacks averaged 5.8 percent higher unemployment than whites under the two presidents before Obama: the rate was 5.0 percent under Bush and reached a 5.5 percent average under Clinton. Under Obama, blacks have had a 6.9 percent higher unemployment rate. Although black unemployment has always been double that of whites, it hasn't usually been 14 percent.

The labor force participation rate offers an even more accurate and damning sketch of black job health because the prolonged denial of opportunity causes many to abandon the quest for jobs in the workforce. The rate in 2009 was 65.7 percent and fell to 62.8 percent in April 2014. For black adults, the figure fell from 63.2 percent to 60.9 percent. When Obama came into office, 29.6 percent of blacks between sixteen and nineteen were working, but by the first quarter of 2014, only 27.9 percent were employed.

When Obama rose to power, 14.3 percent of Americans were trapped beneath the poverty line, a number that rose to 15.0 percent in 2012. Obama often speaks of the middle class but barely mentions the poor, ignoring the one American in seven who falls beneath the poverty line—a dispiriting total of 45 million citizens. Forty percent of black children attend high-poverty schools, while only 6 percent of white students do. When Obama got to the Oval Office, the number of black food stamp recipients stood at 7,393,000, a number that has since risen to 10,955,000. And black home ownership slipped from 46.1 percent at the time Obama took office to 43.3 percent.

The chasm between white and black wealth has increased to 16 to 1. In Obama's first term, the top 1 percent of citizens pulled in 95 percent of all income gains, the worst disparity since a year before the Great Crash on Wall Street in 1929. Neither does Obama's strategic inadvertence successfully offset persistent racial bias in the marketplace. The Justice Department in 2012 struck a $175 million settlement with Wells Fargo & Co., charging that the company had steered approximately four thousand black and Latino borrowers into subprime mortgages, when non-Latino whites with similar credit profiles received prime loans. The company charged approximately thirty thousand minori-

ties higher fees than whites, too. The Justice Department also negotiated the largest-ever fair-lending settlement with Bank of America in
2011 for $335 million. Bank of America's mortgage unit, Countrywide
Financial, had charged blacks and Latinos higher fees and interest rates
than whites with the same credit portfolio, a deceitful practice that Assistant Attorney General Thomas Perez termed "discrimination with
a smile." "Discrimination with a Smile," *New York Times,* January 22,
2015, http://www.nytimes.com/2015/01/22/opinion/discrimination
-with-a-smile.html?_r=0. A 2011 study projected that nearly one-quarter
of black borrowers would have lost their homes to foreclosure by the
end of the crisis; in 2012 foreclosure rates for blacks were nearly twice
as high as those for whites, 9.8 percent versus 5.0 percent. This will have
a deleterious effect on new black wealth creation, since entrepreneurs
routinely tap home equity to finance new ventures. Because the net
worth of black folk is far more likely than that of whites to be tied to
home equity, blacks lost a great deal of equity and wealth when the
housing bubble burst. Before the crisis, whites were already almost
twice as likely as blacks to run their own businesses; the disappearance
of housing wealth leaves blacks without a financial cushion to cover
college tuition or unexpected expenses while depressing the market for
black-owned businesses. Debbie Gruenstein Bocian, Carolina Wei Li,
and Roberto G. Quercia, "Lost Ground, 2011: Disparities in Mortgage
Lending and Foreclosures," Center for Responsible Lending, November
2011, http://www.responsiblelending.org/mortgage-lending/research
-analysis/Lost-Ground-2011.pdf.

16. The Obama administration seemed poised to aggressively reduce racial
segregation of residential neighborhoods after the Supreme Court decision upholding that disparate effect is a legal basis on which to judge
whether a plaintiff is a victim of housing discrimination. As the *New
York Times* reported: "The new rules are an effort to enforce the goals of
the civil rights–era fair housing law that bans overt residential discrimination, but whose broader mandate for communities to actively foster
integration has not been realized. They are part of President Obama's
attempt to address the racial imbalances and lack of opportunity that
he says have contributed to unrest reminiscent of the turbulent 1960s
in cities like Ferguson, Mo., and Baltimore, where African-Americans
have clashed with police officers." Julie Hirschfeld Davis and Binyamin

Appelbaum, "Obama Unveils Stricter Rules on Fair Housing," *New York Times*, July 9, 2015, http://www.nytimes.com/2015/07/09/us/hud -issuing-new-rules-to-fight-segregation.html.

17. O'Reilly, *Nixon's Piano.*

18. "Obama's Father's Day Remarks," *New York Times*, June 15, 2008, http: //www.nytimes.com/2008/06/15/us/politics/15text-obama.html ?pagewanted=all.

19. Martin Luther King, "An Address by Dr. Martin Luther King, Jr.," in *The Moynihan Report and the Politics of Controversy*, ed. Lee Rainwater and William L. Yancy (Cambridge: MIT Press, 1967), p. 407; Michael To-masky, *Left for Dead: The Life, Death, and Possible Resurrection of Progressive Politics in America* (New York: Free Press, 1996), p. 104.

20. Rebekah Levine Coley and Bethany L. Medeiros, "Reciprocal Longitudinal Relations Between Nonresident Father Involvement and Adolescent Delinquency," *Child Development* 878, no. 1 (January–February 2007): 132–147. Also see Daphne C. Hernandez and Rebekah Levine Coley, "Measuring Father Involvement Within Low-Income Families: Who Is a Reliable and Valid Reporter?," *Parenting* 7, no. 1 (2007): 69–97.

21. Eddie Stone, *Jesse Jackson: A Biography* (Los Angeles: Holloway House, 1989), p. 137.

22. Kathleen Hennessey, "In Chicago, Obama Stresses Community, Family in Curbing Violence," *Los Angeles Times*, February 15, 2013, http://articles .latimes.com/2013/feb/15/news/la-pn-obama-chicago-preventing -violence-20130215.

23. *The Melissa Harris-Perry Show*, MSNBC, February 17, 2013, http://www .nbcnews.com/id/50858848/ns/msnbc/t/melissa-harris-perry-show -sunday-february-th/#.Vez_iuvdVJo.

24. "Remarks by the President at Congressional Black Caucus Foundation Annual Phoenix Awards Dinner," September 24, 2011, https://www .whitehouse.gov/the-press-office/2011/09/24/remarks-president -congressional-black-caucus-foundation-annual-phoenix-a.

25. Mackenzie Weinger, "Waters: Obama Remarks 'Curious,'" *Politico*, September 26, 2011, http://www.politico.com/story/2011/09/waters -obama-remarks-curious-064405.

26. "Transcript: Obama's Commencement Speech at Morehouse College," *Wall Street Journal*, May 20, 2013, http://blogs.wsj.com/wash wire/2013/05/20/transcript-obamas-commencement-speech-at-more

house-college/. Contrast Obama's Morehouse speech with his 2012 Barnard College commencement address, in which he praised the female graduates profusely and encouraged them to persevere, in which he empathized with them for the sexism and gender limits they confronted, and in which he committed himself to fighting the problems that plagued them without scolding them in the least: "Of course, as young women, you're also going to grapple with some unique challenges, like whether you'll be able to earn equal pay for equal work; whether you'll be able to balance the demands of your job and your family; whether you'll be able to fully control decisions about your own health. And while opportunities for women have grown exponentially over the last 30 years, as young people, in many ways you have it even tougher than we did . . . I've seen your passion and I've seen your service. I've seen you engage and I've seen you turn out in record numbers. I've heard your voices amplified by creativity and a digital fluency that those of us in older generations can barely comprehend. I've seen a generation eager, impatient even, to step into the rushing waters of history and change its course. And that defiant, can-do spirit is what runs through the veins of American history. It's the lifeblood of all our progress. And it is that spirit which we need your generation to embrace and rekindle right now . . . Indeed, we know we are better off when women are treated fairly and equally in every aspect of American life — whether it's the salary you earn or the health decisions you make . . . And I believe that the women of this generation — that all of you will help lead the way. Now, I recognize that's a cheap applause line when you're giving a commencement at Barnard. It's the easy thing to say. But it's true." "Transcript of Speech by President Barack Obama," Barnard College Commencement, May 14, 2012, https://barnard.edu/headlines/transcript-speech-president -barack-obama.

27. Ta-Nehisi Coates, "How the Obama Administration Talks to Black America," *The Atlantic,* May 20, 2013, http://www.theatlantic.com/politics /archive/2013/05/how-the-obama-administration-talks-to-black -america/276015/.

28. "Remarks by the President at the 'Let Freedom Ring' Ceremony Commemorating the 50th Anniversary of the March on Washington," August 28, 2013, https://www.whitehouse.gov/photos-and-video/video

/2013/08/28/president-obama-marks-50th-anniversary-march
-washington#transcript.

29. Chauncey DeVega, "White America's Racial Amnesia: The Sobering
Truth About Our Country's 'Race Riots,'" *Salon*, May 1, 2015, http://www
.salon.com/2015/05/01/white_americas_racial_amnesia_the_sober
ing_truth_about_our_countrys_race_riots_partner/; "Major Race
Riots in the U.S.," *Infoplease*, http://www.infoplease.com/us/history
/race-riots.html; Sheila Smith McKoy, *When Whites Riot: Writing Race
and Violence in American and South African Cultures* (Madison: University
of Wisconsin Press, 2001).

30. National Advisory Commission on Civil Disorders (the Kerner Report),
1967, http://www.blackpast.org/primary/national-advisory-commis
sion-civil-disorders-kerner-report-1967.

31. Douglas Blackmon, *Slavery by Another Name: The Re-Enslavement of Black
Americans from the Civil War to World War II* (New York: Random House,
2008).

32. Ben Smith, "Obama on Small-Town Pa.: Cling to Religion, Guns,
Xenophobia," *Politico*, April 11, 2008, http://www.politico.com/blogs
/bensmith/0408/Obama_on_smalltown_PA_Clinging_religion
_guns_xenophobia.html.

33. Lisa Bloom, *Suspicion Nation: The Inside Story of the Trayvon Martin Injustice and Why We Continue to Repeat It* (Berkeley: Counterpoint, 2014).

34. Amy Davidson, "If I Had a Son, He'd Look Like Trayvon," *The New
Yorker*, March 23, 2012, http://www.newyorker.com/news/amy-david
son/if-i-had-a-son-hed-look-like-trayvon. Also see Krissah Thompson
and Scott Wilson, "Obama on Trayvon Martin: 'If I Had a Son, He'd
Look Like Trayvon,'" *Washington Post*, March 23, 2012, http://www
.washingtonpost.com/politics/obama-if-i-had-a-son-hed-look-like
-trayvon/2012/03/23/gIQApKPpVS_story.html.

35. Bernard Goldberg, "If President Obama Had a Son He Might Look
Like . . . ," August 22, 2013, http://bernardgoldberg.com/if-president
-obama-had-a-son-he-might-look-like/.

6. Dying to Speak of Race: Policing Black America

1. "Obama: Police Who Arrested Professor 'Acted Stupidly,'" CNN, July
23, 2009, http://www.cnn.com/2009/US/07/22/harvard.gates.inter
view/; "Gates vs. Crowley," *Christian Science Monitor*, July 23, 2009, http:
//www.csmonitor.com/Commentary/the-monitors-view/2009/0723/

po8so1-comv.html; Stephen Brooks, Douglas Koopman, and J. Matthew Wilson, eds., *Understanding American Politics,* 2nd ed. (North York, Ont.: University of Toronto Press, 2013), p. 255.

2. "Obama: Police Who Arrested Professor 'Acted Stupidly.'"

3. "Remarks by President Obama and Prime Minister Abe of Japan in Joint Press Conference," April 28, 2015, https://www.whitehouse.gov /the-press-office/2015/04/28/remarks-president-obama-and-prime -minister-abe-japan-joint-press-confere.

4. Jelani Cobb, "Chronicle of a Riot Foretold," *The New Yorker,* November 25, 2014, http://www.newyorker.com/news/daily-comment/chronicle -ferguson-riot-michael-brown.

5. Transcript, "Case: State of Missouri v. Darren Wilson," Grand Jury vol. 5, September 16, 2014, https://www.documentcloud.org/documents /1370494-grand-jury-volume-5.html; Josh Sanburn, "All the Ways Darren Wilson Described Being Afraid of Michael Brown," *Time,* November 25, 2014, http://time.com/3605346/darren-wilson-michael-brown -demon/; Jake Halpern, "The Cop," *The New Yorker,* August 10, 2015, http://www.newyorker.com/magazine/2015/08/10/the-cop.

6. Ann Petry, *The Street* (New York: Houghton Mifflin, 1946), p. 199.

7. "Statement by the President," Edgartown, Mass., August 14, 2014, https://www.whitehouse.gov/the-press-office/2014/08/14/statement -president.

8. Barack Obama, "Statement by the President," August 18, 2014, https:// www.whitehouse.gov/the-press-office/2014/08/18/statement-presi dent.

9. Barack Obama, "Remarks by the President After Announcement of Decision by the Grand Jury in Ferguson, Missouri," November 24, 2014, https://www.whitehouse.gov/the-press-office/2014/11/24/remarks -president-after-announcement-decision-grand-jury-ferguson-missou.

10. Devin Dwyer, "Obama on Ferguson: 'I Don't Have Any Sympathy' for Protesters Burning Buildings," ABC News, November 25, 2014, http://abc news.go.com/Politics/obama-ferguson-sympathy-protesters-burning -buildings/story?id=27181837.

11. Stephanie Smith, "Bill Bratton Takes Tough Questions from African-American Leaders," *New York Post,* April 4, 2014, http://pagesix.com /2014/04/04/bill-bratton-takes-tough-questions-from-african-ameri can-leaders/.

12. "Giuliani and Dyson Argue over Violence in Black Communities,"

Meet the Press, November 23, 2014, http://www.nbcnews.com/storyline /michael-brown-shooting/giuliani-dyson-argue-over-violence-black -communities-n254431.

13. "Statement by the President" (on Iraq and Ferguson), August 14, 2014, https://www.whitehouse.gov/the-press-office/2014/08/18/statement -president.

14. Michael Eric Dyson, *Is Bill Cosby Right? Or Has the Black Middle Class Lost Its Mind?* (New York: Basic Civitas Books, 2005), p. 142.

15. Kevin Johnson, Meghan Hoyer, and Brad Heath, "Local Police Involved in 400 Killings Per Year," August 15, 2014, http://www.usatoday.com /story/news/nation/2014/08/14/police-killings-data/14060357/.

16. Krissah Thompson, "Rep. John Lewis on 'Selma' and the Memories It Brings to Life," *Washington Post,* December 25, 2014, https://www .washingtonpost.com/lifestyle/style/rep-john-lewis-on-selma-and -the-memories-it-brings-to-life/2014/12/25/f28bab8c-849d-11e4-b9b7 -b8632ae73d25_story.html; "Leaving Selma: A Documentary from the Acclaimed Series 'Andrew Young Presents,'" http://leavingselma.com /timeline.

17. Justin Fenton, "Six Baltimore Police Officers Indicted in Death of Freddie Gray," *Baltimore Sun,* May 21, 2015, http://www.baltimoresun.com /news/maryland/freddie-gray/bs-md-freddie-gray-officer-indictments -20150521-story.html.

18. "Sandtown-Winchester/Harlem Park," Justice Policy Institute, Prison Policy Initiative, February 2105, http://static.prisonpolicy.org/origin/md /Sandtown.pdf.

19. "MLK: A Riot Is the Language of the Unheard," *CBS Reports,* September 27, 1966, http://www.cbsnews.com/news/mlk-a-riot-is-the-language -of-the-unheard/.

20. Naomi Martin, "Parking Dispute May Have Lit Fuse in Waco Biker Shootout," *Dallas Morning News,* May 18, 2015, http://www.dallasnews .com/news/state/headlines/20150518-parking-dispute-may-have -lit-fuse-in-waco-biker-shootout.ece.

21. Reverend Jesse Jackson, "Remarks at the Funeral of Freddie Gray," New Shiloh Baptist Church, Baltimore, April 27, 2015; remarks in possession of the author.

22. "Dajerria Becton: 5 Fast Facts You Need to Know," Heavy.com, June 8, 2015, http://heavy.com/news/2015/06/dajerria-becton-mckinney

-texas-black-girl-bikini-name-assaulted-video-photo-interview-friends
-eric-casebolt/; Doktor Zoom, "Hero Cop Protects Texas from Black
Teenagers at Pool Party," *Wonkette,* June 8, 2015, http://wonkette
.com/587722/hero-cop-protects-texas-from-black-teenagers-at-pool
-party; Brandon Brooks, "Cops Crash Pool Party (Original)," YouTube,
June 6, 2015, www.youtube.com/watch?v=R46-XTqXkzE; "Texas Teen
Girl Interview After Cops Crash Pool Party, Slams Her to Ground, Pulling
Gun on Kids," YouTube, June 8, 2015, https://www.youtube.com/watch
?v=1W2IbpHbopY.

23. "Dajerria Becton: 5 Fast Facts."

24. Kimberlé Williams Crenshaw and Andrea J. Ritchie with Rachel Anspach,
Rachel Gilmer, and Luke Harris, "Say Her Name: Resisting Police Brutality Against Black Women," African American Policy Forum, July 15,
2015, http://static1.squarespace.com/static/53f20d90e4b0b80451158d8c
/t/55a810d7e4b058f342f55873/1437077719984/AAPF_SMN_Brief_full
_singles.compressed.pdf.

25. "Remarks by the First Lady at Tuskegee University Commencement
Address," May 9, 2015, https://www.whitehouse.gov/the-press-office
/2015/05/09/remarks-first-lady-tuskegee-university-commencement
-address.

26. Lis Power, "Right-Wing Media Accuse 'Angry' Michelle Obama of 'Race
Baiting' in Tuskegee Commencement Address," *Media Matters,* May 11,
2015, http://mediamatters.org/research/2015/05/11/right-wing-media
-accuse-angry-michelle-obama-of/203609.

27. Ronald Kessler, *The First Family Detail: Secret Service Agents Reveal the Hidden Lives of the Presidents* (New York: Crown Forum, 2014), p. 41.

28. "Remarks by President Obama and Prime Minister Abe of Japan in Joint
Press Conference."

7. Going *Bulworth:* Black Truth and White Terror in the Age of Obama

1. Peter Baker, "Onset of Woes Casts Pall over Obama's Policy Aspirations," *New York Times,* May 15, 2013, http://www.nytimes.com
/2013/05/16/us/politics/new-controversies-may-undermine-obama
.html.

2. Ezra Klein, "If Obama Went Bulworth, Here's What He'd Say," *Washington Post,* May 16, 2013, http://www.washingtonpost.com/news/

wonkblog/wp/2013/05/16/if-obama-went-bulworth-heres-what-hed
-say/; Melinda Henneberger, "If Obama Did 'Go Bulworth,'" What
Would He Say?," *Washington Post,* May 16, 2013, http://www.washing
tonpost.com/blogs/she-the-people/wp/2013/05/16/obama-feeling
-constricted-longs-to-go-bulworth/. Henneberger quotes Catholic University political scientist Steve Schneck, co-chair of Catholics for Obama
in 2012, who said, "He's done very little for the African American community that went all out for him twice." Henneberger writes, "Schneck
feels sure the community organizer whose idealism got even Washington's hopes up when he first arrived in the Senate is still in there somewhere, 'but he's kept those passions bottled up and out of sight.'"

3. F. Gary Gray's brilliant 2015 biopic *Straight Outta Compton,* the story of
the irreverent eighties and nineties gangsta rap group N.W.A., impres-
sively mines the same fertile political and cultural territory, except it
focuses attention on the plague of police brutality that led to one of
the group's most powerful and controversial hits, "Fuck tha Police,"
and thus establishes its relevance to contemporary black and brown
struggles against police brutality. Lin-Manuel Miranda's remarkable
2015 Broadway play *Hamilton* makes ingenious use of hip hop to explore
vibrant political truths centered in the life and death of Founding Father
Alexander Hamilton, brilliantly fusing the serpentine rhythms of rap
and traditional Broadway melodies to engage complicated ideas about
democracy, race, class, color, and immigration in America.

4. *Bulworth,* DVD, directed by Warren Beatty (1998; 20th Century Fox,
1999).

5. Teresa Tritch, "President Obama Could Unmask Big Political Donors,"
New York Times, March 23, 2015, http://takingnote.blogs.nytimes.com
/2015/03/23/president-obama-could-unmask-big-political-donors/
?_r=0. As Tritch points out, Obama has talked a good game about the
corrupting influence of dark money in politics, so fifty public advocacy
and good government groups sent him a letter (http://www.citizen
.org/documents/2015-sign-on-letter-for-govt-contracting.pdf) challenging him to, in a sense, put his money where his mouth is—or at least to
put his political influence there and sign an executive order "requiring
full disclosure of political spending by corporations receiving federal
contracts, as well as by their directors and officers."

6. "Obama Heckled at White House LGBT Pride Event, Tells Heckler:

'You're in My House!' 'Shame on You!,'" YouTube, June 24, 2015, https://www.youtube.com/watch?v=w71OGC6Jx9w&feature=youtu.be; "President Obama Boots Heckler from White House Event: 'You're in My House,'" *Atlanta Journal-Constitution*, June 24, 2015, http://www.ajc.com/news/news/national/president-obama-boots-heckler-white-house-event-yo/nmkcs/; Justin Wm. Moyer, "Transgender Obama Heckler Jennicet Gutiérrez Hailed by Some LGBT Activists," *Washington Post*, June 26, 2015, http://www.washingtonpost.com/news/morning-mix/wp/2015/06/26/transgender-obama-heckler-hailed-by-some-lgbt-activists/.

7. "Remarks by the President at White House Correspondents' Association Dinner," April 25, 2015, https://www.youtube.com/watch?v=w71OGC6Jx9w&feature=youtu.be; "Obama's Full speech at 2015 White House Correspondents' Dinner," *Washington Post* (*PostTV*), April 25, 2015, http://www.washingtonpost.com/posttv/national/obamas-full-speech-at-2015-white-house-correspondents-dinner/2015/04/25/1ee9a604-ebc1-11e4-8581-633c536add4b_video.html.

8. "Attorney General Eric Holder at the Department of Justice African American History Month Program," February 18, 2009, http://www.justice.gov/opa/speech/attorney-general-eric-holder-department-justice-african-american-history-month-program.

9. Richard Cohen, "Racism vs. Reality," *Washington Post*, July 15, 2013, https://www.washingtonpost.com/opinions/richard-cohen-racism-vs-reality/2013/07/15/4f419eb6-ed7a-11e2-a1f9-ea873b7e0424_story.html.

10. *Shelby County, Alabama v. Holder, Attorney General, et al.*, U.S. Supreme Court (October term, 2012), http://www.supremecourt.gov/opinions/12pdf/12-96_6k47.pdf.

11. Richard Benjamin, "Obama's Safe, Overrated and Airy Speech," *Salon*, July 19, 2013, http://www.salon.com/2013/07/19/obamas_safe_overrated_and_airy_speech/.

12. Danielle Cadet, "Jordan Davis' Shooter Rants About Killing 'Thugs' So They 'May Take the Hint and Change Their Behavior," *Huffington Post*, October 18, 2013, http://www.huffingtonpost.com/2013/10/18/jordan-davis-shooter-michael-dunn_n_4123805.html.

13. Brendan Connor, "Here Is What Appears to Be Dylann Roof's Racist Manifesto," *Gawker*, June 20, 2015, http://gawker.com/here-is-what-appears-to-be-dylann-roofs-racist-manifest-1712767241.

14. "Statement by the President on the Shooting in Charleston, South Carolina," June 18, 2015, https://www.whitehouse.gov/the-press-office/2015/06/18/statement-president-shooting-charleston-south-carolina.

15. Jaeah Lee and Edwin Rios, "Obama to US Mayors on Guns: 'We Need a Change in Attitude. We Have to Fix This,'" *Mother Jones,* June 19, 2015, http://www.motherjones.com/politics/2015/06/obama-mayors-charleston-gun-violence-speech-video.

16. "Hillary Clinton: U.S. Conference of Mayors Speech," San Francisco, June 20, 2015, http://www.shallownation.com/2015/06/20/video-transcript-hillary-clinton-u-s-conference-of-mayors-speech-san-francisco-ca-june-20-2015/.

17. Campbell Robertson, Monica Davey, and Julie Bosman, "Calls to Drop Confederate Emblems Spread Nationwide," *New York Times,* June 23, 2015, http://www.nytimes.com/2015/06/24/us/south-carolina-nikki-haley-confederate-flag.html.

18. David Sims, "The President's Candid Garage Interview," *The Atlantic,* June 22, 2015, http://www.theatlantic.com/entertainment/archive/2015/06/obama-wtf-marc-maron/396488/.

19. Michelle Alexander, *The New Jim Crow: Mass Incarceration in the Age of Colorblindness* (New York: New Press, 2010).

20. Barack Obama, "Remarks by the President at the NAACP Conference," July 14, 2015. Pennsylvania Convention Center, Philadelphia, https://www.whitehouse.gov/the-press-office/2015/07/14/remarks-president-naacp-conference.

21. At a July 2015 press conference on the Iranian nuclear deal, Obama fielded a question from American Urban Radio Networks White House correspondent April Ryan about whether the president would revoke Cosby's 2002 Medal of Freedom. Obama replied:

"With respect to the Medal of Freedom, there's no precedent for revoking a medal. We don't have that mechanism. As you know, I make it a policy not to comment on the specifics of cases where there still might be, if not criminal, then civil issues involved. I'll say this: if you give a woman, or a man, for that matter, without his or her knowledge, a drug, and then have sex with that person without consent, that's rape. And I think this country, or any civilized country, should have no tolerance for rape." See Kia Makarechi, "Obama's Striking Response to Bill Cosby Rape Allegations," *Vanity Fair,* July 15, 2015, http://www.vanityfair.com/hollywood/2015/07/obama-bill-cosby.

22. See Michael Eric Dyson, *Come Hell or High Water: Hurricane Katrina and the Color of Disaster* (New York: Basic Civitas Books, 2005), p. 20; Bruce Alpert, "Transcript of President Obama's Katrina Speech," NOLA.com /*Times-Picayune*, August 28, 2015, http://www.nola.com/katrina/index .ssf/2015/08/transcript_of_president_obamas.html.

23. Alan Borass, "Obama's Decision on Denali Strikes a Blow for Decolonization and Respect," *Alaska Dispatch News,* September 6, 2015, http:// www.adn.com/article/20150906/obamas-decision-denali-strikes -blow-decolonization-and-respect.

24. David Smith, "12 Things We Learned from Obama's Historic Trip to Africa," *The Guardian*, July 31, 2015, http://www.businessinsider.com /obama-trip-to-africa-kenya-ethiopia-2015-7.

8. Amazing Grace: Obama's African American Theology

1. Chris Cillizza, "This Was the Best Week of Obama's Presidency," *Washington Post,* June 26, 2015, http://www.washingtonpost.com/blogs/the -fix/wp/2015/06/26/this-was-the-best-week-of-obamas-presidency/; Eugene Scott, "Obama Describes 'Best Week Ever,'" CNN, June 30, 2015, http://www.cnn.com/2015/06/30/politics/obama-best-week -ever-press-conference/; Nick Gass, "Barack Obama's Best Week Ever (It Wasn't Last Week)," *Politico,* June 30, 2015, http://www.politico .com/story/2015/06/obama-best-week-ever-marry-michelle-119594 .html. In typically humorous and self-deprecating fashion, Obama, while acknowledging the historic character of the seven-day stretch, named three superior candidates for the best weeks of his life: when he married wife Michelle Obama, when his daughters Sasha and Malia were born, and when he scored twenty-seven points in a basketball game. For a fascinating portrait of Obama as hoopsman—as a figure whose identity basketball helped to shape, the same game that permitted him to impress his future wife and define himself as a political leader—see Alexander Wolff, *The Audacity of Hoop: Basketball and the Age of Obama* (Philadelphia: Temple University Press, 2015).

2. Less than a week after Dylann Roof, on June 17, killed the nine members of the Emanuel AME Church in Charleston—known as the Emanuel 9—Governor Nikki Haley called, on June 22, for the South Carolina state legislature to remove the Confederate flag from its grounds; a little more than two weeks later, on July 9, she signed legislation for the flag to be removed, and it came down on July 10, 2015, at a

little after 10:00 a.m. Eastern Daylight Time. In calling for the flag to be taken down, Haley made an impassioned speech in which she stated, in part, that for many South Carolinians, "the flag stands for traditions that are noble. Traditions of history, of heritage, and of ancestry.

"The hate filled murderer who massacred our brothers and sisters in Charleston has a sick and twisted view of the flag. In no way does he reflect the people in our state who respect and, in many ways, revere it. Those South Carolinians view the flag as a symbol of respect, integrity, and duty. They also see it as a memorial, a way to honor ancestors who came to the service of their state during time of conflict. That is not hate, nor is it racism.

"At the same time, for many others in South Carolina, the flag is a deeply offensive symbol of a brutally oppressive past. As a state we can survive, as we have done, while still being home to both of those viewpoints. We do not need to declare a winner and a loser here. We respect freedom of expression, and that for those who wish to show their respect for the flag on their private property, no one will stand in your way.

"But the statehouse is different and the events of this past week call upon us to look at this in a different way . . .

"But this is a moment in which we can say that that flag, while an integral part of our past, does not represent the future of our great state. The murderer now locked up in Charleston said he hoped his actions would start a race war. We have an opportunity to show that not only was he wrong, but that just the opposite is happening.

"My hope is that by removing a symbol that divides us, we can move forward as a state in harmony and we can honor the nine blessed souls who are now in heaven." "Transcript: Gov. Nikki Haley of South Carolina on Removing the Confederate Flag," *New York Times,* June 22, 2015, http://www.nytimes.com/interactive/2015/06/22/us/Transcript -Gov-Nikki-R-Haley-of-South-Carolina-Addresses-Removing-the-Con federate-Battle-Flag.html. On July 10, 2015, the flag finally came down. See Richard Fausset and Alan Blinder, "Era Ends as South Carolina Lowers Confederate Flag," *New York Times,* July 10, 2015, http://www.ny times.com/2015/07/11/us/south-carolina-confederate-flag.html.

3. Hans A. Baer and Merrill Singer, *African American Religion: Varieties of Protest and Accommodation,* 2nd ed. (Knoxville: University of Tennessee

Press, 2002), pp. 10–11 [first edition published under the title *African-American Religion in the Twentieth Century: Varieties of Protest and Accommodation*]; Cornel West, *The Cornel West Reader* (New York: Basic Civitas Books, 1999), pp. 62–63.

4. Barack Obama, "Remarks by the President in Eulogy for the Honorable Reverend Clementa Pinckney," College of Charleston, Charleston, S.C., June 26, 2015, https://www.whitehouse.gov/the-press-office/2015/06/26/remarks-president-eulogy-honorable-reverend-clementa-pinckney. For a brilliant reading of Obama's eulogy, one that explores "Obama's gifts of language and empathy and searching intellect," and the use of those gifts "to talk about the complexities of race and justice, situating them within an echoing continuum in time that reflected both Mr. Obama's own long view of history, and the panoramic vision of America, shared by Abraham Lincoln and the Rev. Dr. Martin Luther King Jr., as a country in the process of perfecting itself," see Michiko Kakutani, "Obama's Eulogy, Which Found Its Place in History," *New York Times*, July 3, 2015, http://www.nytimes.com/2015/07/04/arts/obamas-eulogy-which-found-its-place-in-history.html. For another brilliant reading of Obama's eulogy, one which says that by "singing a spontaneous congregational song at the end of a sermon — the traditional emotional apex of the ritualized event — Obama performed a familiar trope that united his immediate audience — mostly black church-goers — with their history, that joined himself to that history, and that staged social solidarity among the musicians and the singing congregation," see Guthrie Ramsey, "Obama's 'Amazing Grace' Is a Sign of Music's Powerful Role in Black Churches," *The Guardian*, June 30, 2015, http://www.theguardian.com/commentisfree/2015/jun/30/obama-amazing-grace-charleston-eulogy.

5. Ira Berlin, *The Making of African America: The Four Great Migrations* (New York: Viking, 2010); Peter A. Coclanis, *The Shadow of a Dream: Economic Life and Death in the South Carolina Low Country, 1670–1920* (New York: Oxford University Press, 1989); Peter H. Wood, *Black Majority: Negroes in Colonial South Carolina from 1670 Through the Stono Rebellion* (New York: W. W. Norton, 1996); Bernard E. Powers, *Black Charlestonians: A Social History, 1822–1885* (Fayetteville: University of Arkansas Press, 1999); Steve Estes, *Charleston in Black and White: Race and Power in the South After the Civil Rights Movement* (Chapel Hill: University of North Carolina Press,

2015). Thanks to Henry Louis Gates Jr. for suggesting that I probe this historical dimension of Charleston.

6. Peter H. Wood, "'More Like a Negro Country': Demographic Patterns in Colonial South Carolina, 1700–1740," in *Race and Slavery in the Western Hemisphere: Quantitative Studies*, ed. Stanley L. Engerman and Eugene D. Genovese (Princeton: Princeton University Press, 1975), pp. 131–145.

7. Barack Obama, "Remarks by the President at a Campaign Event in Roanoke, Virginia," July 13, 2012, https://www.whitehouse.gov/the-press -office/2012/07/13/remarks-president-campaign-event-roanoke -virginia.

8. See Libby Nelson, "The Confederate Flag Symbolizes White Supremacy—and It Always Has," *Vox Identities*, June 22, 2015, http://www.vox .com/2015/6/20/8818093/confederate-flag-south-carolina-charleston -shooting.

9. Dietrich Bonhoeffer, *The Cost of Discipleship* (1949; repr., New York: Touchstone, 1995), pp. 44, 45.

10. When President Obama was on the *Marine One* helicopter flying from the White House to Andrews Air Force Base to take his flight to South Carolina for the funeral, he mentioned to White House senior adviser Valerie Jarrett and Mrs. Obama that he was thinking of breaking into song. "When I get to the second part of referring to 'Amazing Grace,' I think I might sing," Jarrett recalls Obama saying to them. Her response was "Hmm," while Michelle Obama pointedly asked, "Why on earth would that fit in?" President Obama replied: "I don't know whether I'm going to do it, but I just wanted to warn you two that I might sing . . . We'll see how it feels at the time." Jarrett later said that she and the first lady eventually "both encouraged him to do whatever the spirit moved him to do." After Obama sang and made a hugely positive impression, Jarrett wondered about the pauses. "So later I said to him, 'Were you thinking about whether or not to sing?' He said, 'Oh no, I knew I was going to sing. I was just trying to figure out which key to sing it [in].'" Jarrett relayed the story at the 2015 Aspen Ideas Festival in a session with Walter Isaacson, the Aspen Institute president. See Peter Baker, "When the President Decided to Sing 'Amazing Grace,'" *New York Times*, July 6, 2015, http://www.nytimes.com/politics/first-draft/2015/07/06/obama baker/?_r=0; and Chuck Ross, "Valerie Jarrett Says Michelle Obama Was Skeptical of Husband's Plan to Sing 'Amazing Grace,'" *Daily Caller*, July 7, 2015, http://dailycaller.com/2015/07/07/valerie-jarrett-says

-michelle-obama-was-skeptical-of-husbands-plan-to-sing-amazing
-grace-video/.

11. As Liam Viney argues: "Obama sings a bit flat. Naturally, as a singer without formal training, who has had certain other things to attend to in recent years, he may have just not had the best singing technique. Intonation insecurity and dubiously executed melismata were balanced by an undeniable connection to African American musical culture. That flatness was very likely Obama channeling the blues." Liam Viney, "Obama's Amazing Grace Shows How Music Can Lift Oratory High," *The Conversation*, June 30, 2015, http://theconversation.com/obamas -amazing-grace-shows-how-music-can-lift-oratory-high-44076.

12. See Albert Raboteau's fine essay "The Chanted Sermon" in his book *A Fire in the Bones: Reflections on African American History* (Boston: Beacon Press, 1995), pp. 141–151.

★ |||||| ★

INDEX